mer

to: two of the
sweetest people ive
ever met. No wonder
such a lovely human
came from you!

Hopefully you
can spot a ~~these~~
few of these trees
in your travels around
BC. lots of love ♡
fenny.

VANCOUVER TREE BOOK

Preface:
The Reason for This Book

This book comes with a warning. Two warnings, actually.

First, you should know that learning about trees may seem fun at first but can quickly grow into an obsession—although the side effects are mostly good.

Don't be surprised to find your relationship change with the city you thought you knew. The urban space you may have brain-mapped as a traffic grid or a collection of neighbourhoods will become a forest—an urban forest, one of the most complex and interesting ecosystems on the planet. Now even your outlook on life may broaden as your view of the living world comes to include the city itself.

To learn trees is to realize a particular specimen you may have ignored or seen as a green backdrop is actually a living being, one like us with stages of life from birth to youth to maturity and decline. Also like us it's thriving or struggling with all the challenges of trying to get ahead in a crowded urban environment.

Look deeper and that tree may turn into an object of perennial wonder, even a friend. The things you can learn about this friend are endless, and they keep changing as the tree goes through seasons of life together with its allies and enemies in the urban forest. So you should know the door you're opening with this book may lead to an outdoor library with way more information than you can learn in this lifetime, and your quest to absorb it all could well be endless.

The second warning is less a caveat than a fair disclosure. This book has a purpose beyond entertaining you with tree stories. It is written in the belief that knowledge is power, and the power of citizens who know the value of the urban forest is needed now more than ever.

We stand at a critical stage in the campaign to stop global climate madness. Time is short and politics are often slow, but we can create a sea change when enough people stand together to say the same thing. If learning more about the most important plants on our living Earth can help bystanders become engaged ecologists, people who realize that our struggle for the planet must be won in the cities where most of us live, whatever effort went into this book finding its way into your hands will be well worth it.

Introduction:
What's In a Name?

The first step in knowing someone is to learn his or her name. It's the same with trees.

When we learn the name of a tree, we recognize it as distinct. In one sense, we may now see it for the first time. Knowing how to distinguish a tree by name lets you see trees as our ancestors once did. These are fellow beings sharing our space. Some may be key to our survival, others so noble they instill reverence.

Why venerate something so common as a tree? The reasons are many, but here are three:

- trees have been the Earth's stewards far longer than us—some go back million of years, even before the dinosaurs—and trees may well be here after we're gone.
- trees are the kings and queens of the plant world, living energy factories that eat sunshine and emit oxygen.
- trees filter polluted air, cool hot cities, reduce stormwater runoff and provide food, beauty and grace that soften the hard edges of urban development.

This field guide was written to help you recognize more than 100 types of Vancouver's most important trees. "Important" is relative; in this case it means species native to here or else widely planted, although it also has room for trees with cultural value or historical significance. Some merit attention just for being remarkable even if rare. With added information on related species and cultivars (short for "cultivated varieties" grown for desired traits), well over 200 trees are described, making this book a handy resource for tree enthusiasts of all levels from beginner to professional arborist.

People and buildings and trees, like these *Platanus x acerifolia*—London plane, can all get along in a healthy urban forest.

How to Use
This Field Guide

Start with a leaf to determine which of nine groups this book uses to separate trees. A leaf can be any shape including needle-like or scaled. The back cover flap has descriptions with photos of the nine groups for easy flipping.

Once you determine the group, turn to that section in the book to find which tree makes the best match. Detailed descriptions and images cover the size, shape, bark, flowers, fruit and more for each tree. Consider the variety of clues in making a call, and don't be dismayed if not everything fits perfectly, especially with some images. Trees are social beings that like to grow together but they're also individuals, and even on the same tree variations may be seen in the size, shape and colour of leaves. A photo can cover only one view of a particular leaf in a particular light.

If you're trying to identify a tree that seems close to a description in the book yet annoyingly insists on putting out contrary features, it could be a related species or a cultivar of the same species. To include details and images of all the thousands of cultivars now available would result in a book so big you would need help carrying it. Instead the strategy here is to help you get as close as possible to nailing the genus (a group of similar types, such as oaks) and then the species (the individual type within the

3

genus, such as red oak). The closer you get with this field guide, the easier your further research will be through the Internet or taxonomy books.

Trees in this book are first divided into two types. Groups 1 to 3 describe conifers (cone-bearing), which are usually but not always evergreen. Groups 4 to 9 describe broadleaf trees, which are usually (but not always) deciduous, meaning they shed their leaves for winter.

The individual groups are then based on descriptions of the leaves. Examine a leaf to figure out which group its tree is in. With conifers, decide if you're looking at leaves/needles in the shape of overlapping scales (Group 1), single needles along the twig (Group 2) or needles in bundles or tufts (Group 3).

For broadleaf trees, check the shape of the leaf to determine whether it's simple (one leaf blade per stalk) or compound (several leaflets per stalk). Simple leaves are further divided according to whether they're shaped like a fan, which makes them ginkgo (*Gingko biloba*), the only tree in Group 4, or not.

If not, see whether they're arranged opposite each other along the twig in pairs (Group 5) or alternately. One mnemonic to remember trees with opposite leaves is "mad horses." It describes **M**aple, **A**sh, **D**ogwood and **H**orsechestnut. If it isn't confusing you can expand it to "mad horses and cats" and add **Kat**sura (and if you include leaves with opposite or whorled arrangements, **C**atalpa.)

Alternate-leaf trees in which the leaf edges have lobes, or wide indentations like you would expect on oaks, make up Group 6. Alternate-leaf trees that have toothed leaf edges such as alders and birches make up Group 7. Group 8 consists of alternate-leaf trees that have leaves with smooth edges such as magnolias. Finally, Group 9 is for compound leaf trees, meaning those with leaflets on a single stalk, and includes walnuts, sumac and the tree of heaven.

Trees within each group are listed alphabetically by their scientific name, typically a binomial (two words) describing the genus and the species and perhaps sounding exotic enough to trip the tongue. Nevertheless, they're worth committing to memory. Common names vary in different regions or even among different people in the same region, but scientific names are universal. Tree enthusiasts the world over can talk to each other using scientific names even if they lack a common language. Also, knowing a scientific name will help make you sound and feel like a true tree geek—yes, that is a good thing.

Botanical classifications offer a roadmap to navigate the urban forest, but you needn't get too bogged down by details. In some cases you may be content to figure out just the genus. This puts related types into the same group. Maples, for example, are all in the genus *Acer*. A genus group is made of a number of species, or specific trees. There are more than 100 kinds of specific maples in the *Acer* genus, including *Acer saccharum* (sugar maple) and *Acer palmatum* (Japanese maple). The oak genus, *Quercus*, has more than 600 species. (You will please the botanical-minded if you remember to write these scientific names according to protocol—genus starting with a capital letter, species starting with a lower case letter, and both in italics.)

Learning to recognize trees does take effort. You need to commit names to memory, which means spending time with trees and this guide, going back and forth from one to the other. If that sounds suspiciously like school, it can be seen that way, but in a positive light. Most consider this quality time. The more opportunities you take to look at a tree, touch the bark, smell the leaves and appreciate the form in the same way you might linger before an art museum Cezanne, the more you'll get out of it. Before long you will come to a point where things just click and you know that tree not by deciphering its parts but because you recognize it instantly, the same way you recognize a good friend.

To hone your tree ID skills, Vancouver has two fine botanical gardens. Their collections are excellent and professionally tended, and most of the trees have labels. If only more city trees could be labelled! Both botanical gardens are listed in the Tree Tour maps section, along with eight other tree-centric sites. The other sites won't have labels on the trees, but the maps list several noteworthy specimens at each place to get you going. Beyond the maps, all profiled trees include locations to find at least one, not necessarily the tree photographed.

Also included is a map of 10 Treasured Trees located throughout the city. This list is clearly subjective, based only on a desire to find specimens interesting for their shape or story, the kind of treasures one might show a tree geek visiting from abroad. Here's hoping you can see them all to understand why. Or even better, to discover Treasured Trees of your own and share with others. Just as no tree is alone in the urban forest, people who love trees should feel encouraged to talk to each other about their discoveries, which could be the start of a wider campaign to protect them.

Lastly, although you will not be advised against it, this book will never ask you hug a tree.

You will however get this advice, passed down from a First Nations elder. He said it while we were resting beside a western redcedar that would have been huge when Christopher Columbus was still a baby. It was a hot afternoon and our group of sweaty hikers was trying to enjoy a shady break when a cloud of mosquitoes descended. We started slapping our arms and necks, usually a second too late. The elder nudged my elbow to demonstrate another approach. He slowly waved a bough of the fragrant redcedar over his bare arms, and no mosquitoes could land. When I thanked him, he offered another lesson. He said you can get strength from a great tree by standing or sitting with your back against the trunk. That way the power of the tree flows into you.

Try it yourself to see if it works.

How We Got Here:
A History of Vancouver Trees

Vancouver is a rainforest, even when it doesn't look like one.

In ecological terms it's a temperate rainforest in the coastal western hemlock biogeoclimatic zone. You can discover what that means in a walk downtown to Stanley Park.

Head in to the forested heart where conifers soar above sword ferns and native berries and salal, and everything is covered by a soft, green carpet of moss. This is not "virgin" or "original" or "old growth" forest—Stanley Park was selectively logged from 1860 to 1880—but it does give you an idea of how local conditions once crafted the rainforest and might yet shape any land left to grow.

Trees tell the story of the city. Historically they provided the Musqueam, Squamish and Tsleil-Waututh rainforest dwellers and seasonal visitors with homes, food, medicine, clothes, tools and more. First Nations people lived in and on the rainforest, granting the trees on which they based their survival a respect equal to that for salmon or game animals or each other. Trees were considered co-inhabitants of a shared space, and in some cases sacred.

When early Europeans saw this rainforest they were awed by the massive height and girth of the trees, some of the largest ever to be found on the planet. Then they cut them down.

Vancouver was built to sell its forest. Even before being declared a city in 1886 it was doing a roaring global trade in trees.

Douglas-fir (*Pseudotsuga menziesii*) provided durable, knot-free wood prized for masts and spars on sailing ships, especially to the Royal Navy. Some were delivered to China to shore up the emperor's Forbidden City. A monument at the foot of Dunlevy Street to Stamp's Mill, one of three in the area chewing through the forest, reads: "On July 25, 1867, the vessel Siam left Stamp's Mill with lumber for Australia, thus beginning Vancouver's prime function: to supply her great timbers to the world."

The British Columbian newspaper reported on June 17, 1868, that in the year since that first shipment, Stamp's Mill exported 4 million feet of lumber, 100,000 shingles and 2,000 spars. Forests that had taken more than a millennium to mature while

Real estate promotion on Granville St, Harry T. Devine photo, 1886, City of Vancouver Archives, CVA 371-2568.

supporting a traditional way of life for generations were reduced to stumps in a few years.

Not all of the early settlers' decisions were as destructive. The first move of the first city council in 1886 was to protect a 1,000-acre forested area from further logging or development—now one of the world's most successful urban green spaces known as Stanley Park.

In 1890 Vancouver created a park board to manage parks and trees with leadership positions filled by election, a rarity in North America. Park board arborists now plant and tend the city's 140,000 street trees as well as those in parks. There are more trees on private land, but in August 2014 when it was understood the urban forest was shrinking, city council amended an earlier bylaw so that residents would no longer be allowed to cut down any tree with a trunk thicker than 20 cm in diameter, even on their own property. This law based on ecology rather than economics recognizes every tree as important to the urban forest, and the forest as a resource shared by all. More cities are passing laws like it every year.

The trees you see today reflect the blend of cultures which make Vancouver unique. Some people wonder whether the urban forest in some places is "natural." Not in the sense they may think, if they're out to champion the cause of native trees. There are excellent reasons for people to plant native species in the city—they tend to do well in our climate (although poor urban soils may leave some struggling), they support native birds and insects and are a visual reminder of where on the planet we are. On the other hand, an entirely native forest would be dominated by large and shady trees, even in winter, and many would miss the sunshine. That doesn't mean they shouldn't be included where they best fit. To see real majesty in wood and leaf, check out some of the few remaining old growth trees in

Vancouver such as in Stanley Park and on the Wreck Beach cliffs off UBC.

Vancouver continues to change but European influences are still strong. Many of the older trees planted in the last 100 years reflect the English fondness for deciduous icons such as oak and ash. But each spring the city now turns into a massive urban bouquet as thousands of flowering Japanese cherries and plums fill the air with aromatic pink and white blossoms.

Although many people love the look of a street lined with all the same type of tree, ecologists recognize the problem with monocultures, among them being the risk of creating a highway for pathogens that can destroy the whole lot. Many of the flowering cherries planted so profusely in the last 40 or 50 years suffer from diseases that thrive in gloppy mild winters and springs, so the city has since adopted diversity as a guide for its species selection. The goal for each of Vancouver's 22 neighbourhoods is to have no more than 20 percent of the trees from the same genus, and no more than 10 percent from the same species. This is good policy from the environmental view where diversity equals strength.

Vancouver launched a Greenest City campaign in 2011 which included the goal to add 150,000 trees by the year 2020. Recognizing the limited open space available, a call went out for citizens to join the effort by planting more trees around their homes.

The result of all this is a rich blend of species from around the world, adding up to an urban forest unique to this place and time and one on which the city relies for much of its famed beauty. The same growing conditions that led to the rainforest behemoths are now supporting a variety of trees as eclectic and dynamic as the city itself. Although climate change may mean all bets are off when it comes to growing conditions, the typical weather here with wet, mild winters and warm but not too hot summers creates a paradise for trees.

It's also an excellent place to see and study them. Visiting tree experts, especially those from back east, are invariably impressed by the diversity of the urban forest in Vancouver. They may not however know its challenges: the canopy cover (percentage of land covered by trees when viewed from above), which would have been nearly complete before European contact, is now just 18 percent. That's low for a city in North America, particularly for a city that trades so highly on its natural beauty. It recent years the percentage has been steadily dropping, mostly due to development.

The challenge for Vancouver's future will be growing an urban forest that can live up to the legacy of its past while accommodating the many who want to live here in the future. The only way to get this right is to move forward while looking back, honouring and continuing to learn from those who understand the rainforest as home. City council took a step in this direction on June 25, 2014, when it unanimously recognized the "truth that the modern city of Vancouver was founded on the traditional territories of the Musqueam, Squamish and Tsleil-Waututh First Nations and that these territories were never ceded through treaty, war or surrender."

Building on a rich cultural tradition of living with trees would give Vancouver the chance to nuture an urban forest of world class beauty and strength.

Tree Tour Maps

These maps are offered to whet your sense of adventure while helping you find and identify some fantastic trees. They are not meant to cover every tree in a given space, a task beyond a book this size. For a pleasant educational outing, start with the numbered trees here. You can then add to your knowledge base by exploring nearby trees, some of which may be the same species as mapped trees. Note that map scales vary and tree placements are approximate. Also, trees are living, growing and dying beings, so some entries included at the time of publication may now live only in memory.

Trout Lake—John Hendry Park
3300 Victoria Dr @ E 15th Ave

A busy but friendly East Van choice for a nature break any time of year. The lake itself, a peat bog topped up with city water, is swimmable in summer. The surrounding trees are a curious mix of mostly introduced species planted in different eras. Some are now sizeable and all contribute to a restorative stroll around the lake, even as you negotiate your way through the dogs. The golden catalpa, No. 2, was planted on Sept 27, 2014, as the Vancouver Poet-Tree to inspire free expression.

1 *Fraxinus latifolia*—Oregon ash

2 *Catalpa bignoniodes* 'Aurea'—golden catalpa

3 *Quercus rubra*—red oak

4 *Acer rubra*—red maple

5 *Alnus rubra*—red alder

6 *Populus nigra* 'Italica'—Lombardy poplar

7 *Ulmus carpinifolia*—smoothleaf elm

8 *Salix x sepulcralis* 'Chrysocoma'—golden weeping willow

9 *Cedrus deodara*—deodar cedar

10 *Quercus garryana*—Garry oak

Pacific Central Rail Station—Thornton Park
1166 Main St @ Terminal Ave

A small but stately setting in an interesting neighbourhood, Thornton Park marks the southern entrance to the Downtown Eastside. Its trees, some huge, stand up well against the Neoclassical Revival station building completed for the Canadian Northern Railway in 1919. Construction of the park began in 1923 on a formal geometric design. Every spring the city arboriculture department gets calls from people who insist on knowing the name of that tree, you know, the one with the flowers by the station—and before the caller can gush any further the answer is given: it's a princess tree in Thornton Park.

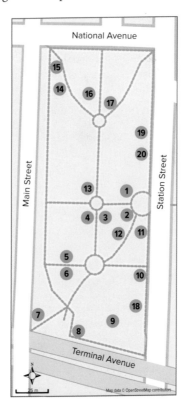

1. *Paulownia tomentosa*—princess tree
2. *Ailanthus altissima*—tree of heaven
3. *Fagus sylvatica*—European beech
4. *Quercus palustris*—pin oak
5. *Catalpa speciosa*—northern catalpa
6. *Juglans nigra*—black walnut
7. *Pterocarya stenoptera*—Chinese wingnut
8. *Pinus strobus*—white pine
9. *Sequoiadendron giganteum*—giant sequoia
10. *Cornus florida*—flowering dogwood
11. *Aesculus x carnea* 'Briotii'—Briotii red horsechestnut
12. *x Chitalpa tashkentensis*—chitalpa
13. *Cercidiphyllum japonicum* 'Pendula'—weeping katsura
14. *Sequoia sempervirens*—redwood
15. *Catalpa bignonioides*—southern catalpa
16. *Quercus rubra*—red oak
17. *Ulmus americana*—American elm
18. *Ilex aquifolium*—English ivy
19. *Acer cappadocicum*—cappadocicum maple
20. *Prunus avium* 'Plena'—double flowered mazzard cherry

Stanley Park

On this tour we're looking for giants. Not all are "old growth," but even some of the 20th century youngsters are awesome displays of beauty and bulk.

History and ecology are rich in what TripAdvisor has rated the "best park in the world." Look for notches in the massive stumps—they held the springboards loggers stood on to cut the tree. You might also find "culturally modified trees," or CMTs, missing sections of bark. First Nations forestry includes techniques to harvest from a tree without killing it.

The first three planted on the lawn west of the golf course gate are a California trio of future hulks. No. 4 on a trail east of the junction of Lees Trail and the Bridle Path is old growth, the largest Douglas-fir in the park. No. 5 off Tatlow Walk southeast of the intersection at Rawlings Trail is the park's widest western redcedar thanks to a huge swollen base. No. 6 east of Rawlings Trail south of the Hollow Tree is Canada's biggest maple tree—a short trail to it begins at a fellow bigleaf maple. The Hollow Tree (No. 7) is no longer a tree at all but the remnants of a photographic icon. Head back south to

see No. 8, a looming grand fir beside Rawlings Trail. Across Park Road on the east side of Merilees Trail No. 9 marks two big Sitka spruce just north of Third Beach. No. 10 north of the concession stand at Third Beach is listed on the BC Big Tree registry as the second largest red alder in the province.

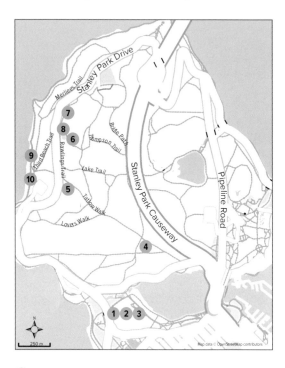

1. *Sequoiadendron giganteum*—giant sequoia
2. *Sequoia sempervirens*—redwood
3. *Calocedrus decurrens*—incense cedar
4. *Pseudotsuga menziesii*—Douglas-fir
5. *Thuja plicata*—western redcedar
6. *Acer macrophyllum*—bigleaf maple
7. The Hollow Tree
8. *Abies grandis*—grand fir
9. *Picea sitchensis*—Sitka spruce
10. *Alnus rubra*—red alder

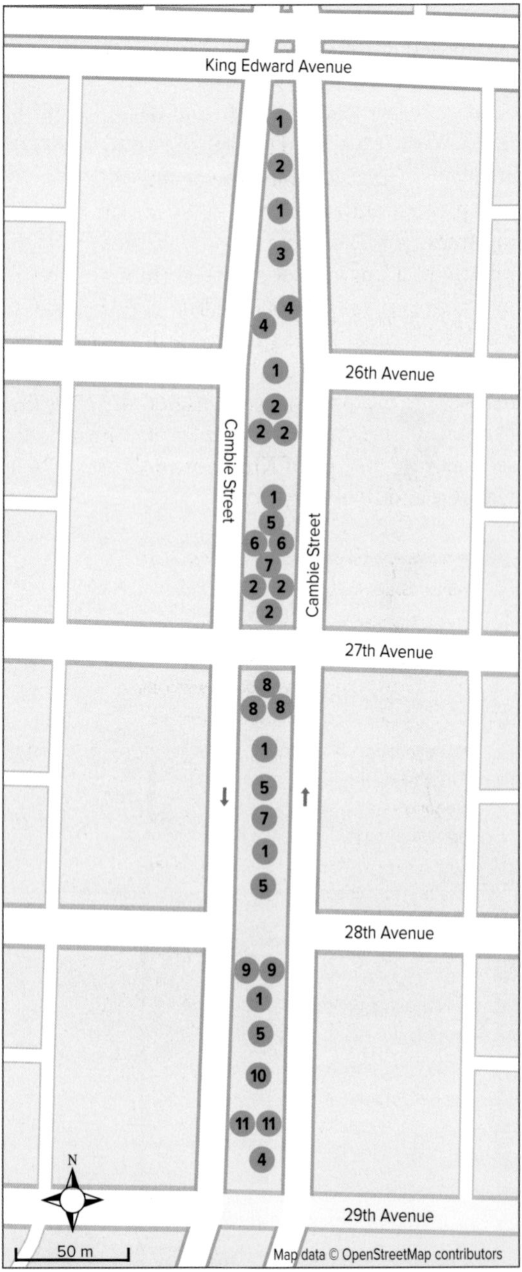

King Edward Avenue

26th Avenue

Cambie Street

Cambie Street

27th Avenue

28th Avenue

29th Avenue

N

50 m

Map data © OpenStreetMap contributors

Cambie Heritage Blvd
Cambie Blvd from King Edward Ave to 29th Ave

Cambie Street is an arterial road designed in the 1920s, an age when road transportation could be still considered pleasureable. The 10-m-wide median was planted with hundreds of shade and flowering trees. Somehow this garden/arboretum in the middle of a busy street survived the pressures of hyper-development and even the construction of a Skytrain line. Rather than an aerial as the name suggests the line was built into an underground tunnel, saving what in 1993 was declared the city's first heritage landscape. Other median plantings can be seen along portions of King Edward Ave, 1st Ave, 16th Ave and Boundary Rd.

1. *Sequoiadendron giganteum*—giant sequoia
2. *Prunus* 'Shirofugen'—Shirofugen cherry
3. *Prunus* 'Accolade'—Accolade cherry
4. *Cedrus deodara*—deodar cedar
5. *Ulmus glabra* 'Lutescens'—golden wych elm
6. *Prunus* 'Snow Goose'—Snow Goose cherry
7. *Picea pungens*—blue spruce
8. *Prunus yedoensis* 'Akebono'—Akebono cherry
9. *Fagus sylvatica* 'Red Obelisk'—Red Obelisk European beech
10. *Abies concolor*—white fir
11. *Fagus sylvatica*—European beech

Strathcona Community Garden
759 Malkin Ave

Neighbours who lacked the space to grow their own organic food turned this unofficial dump into a community garden in 1985, a harbinger of the city's 21st century support for urban agriculture. Today it's seen as a centre of excellence for volunteer-run environmental stewardship, and receives visitors from around the world. The community orchard, recognized by the Vancouver Heritage Foundation, includes more than 300 kinds of fruit trees, which is the theme of this tour. The espalier section has many labeled heritage apples, unique varieties that are clones of their original and have survived only because growers have extended the legacy for generations by grafting.

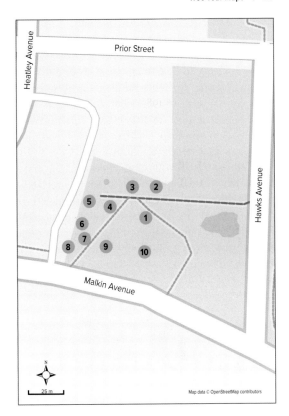

1. *Malus* 'Dolgo'—Dolgo crabapple
2. *Ficus carica*—fig
3. *Malus domestica* 'Drap d'Or'—Drap d'Or apple
4. *Prunus domestica* 'Early Italian'—Early Italian plum
5. *Diospyros kaki* 'Fuyu'—Fuyu persimmon
6. *Pyrus pyrifolia*—Asian pear
7. *Asimina triloba* 'Davis'—Davis pawpaw
8. *Prunus avian* 'Van'—Van cherry
9. *Morus nigra*—black mulberry
10. *Cydonia oblonga*—quince

UBC Botanical Garden
6804 SW Marine Dr

It's worth the trip to UBC to stroll around this research/teaching garden established in 1916. It has trees from around the world, including rare specimens collected in remote places you probably couldn't find on a map. The many labels identifying trees in a variety of collections such as the Asian Garden, Native Garden and Food Garden make this a tree geek wonderland. No. 10 is a Douglas-fir perhaps 600 years old and known as the Eagle Tree for the raptors who like to perch there. For an additional fee you can add to your visit by going on the Greenheart Canopy Walkway, a series of suspended bridges.

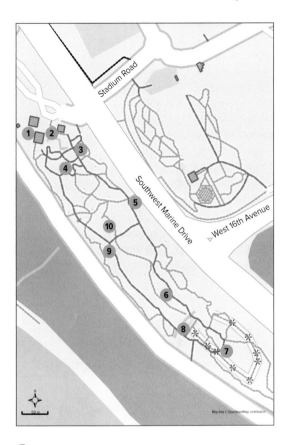

1. *Tapiscia sinensis*—false pistache
2. *x Chitalpa tashkentensis* 'Pink Dawn'—Pink Dawn chitalpa
3. *Carpinus fangiana*—Fang's hornbeam
4. *Cercidiphyllum japonicum* 'Morioka weeping'—weeping katsura
5. *Stewartia pseudocamellia*—Japanese stewartia
6. *Trochodendron araliodes*—wheel tree
7. *Bretschneidera sinensis*—bretschneidera
8. *Nyssa sinensis*—Chinese tupelo
9. *Acer griseum*—paperbark maple
10. *Pseudotsuga menziesii*—Douglas-fir

VanDusen Botanical Garden
5251 Oak St

Planted in 1975 on the grounds of an old golf course, this 22-ha space merits repeat visits, not because you can't cover it all in a few hours but because you'll want to return anyway. It's probably the best site in the city to learn tree ID in a hurry as the many labels reveal the correct answers. It's also a fine place for a green-minded stroll through various collections that include both familiar and rare trees.

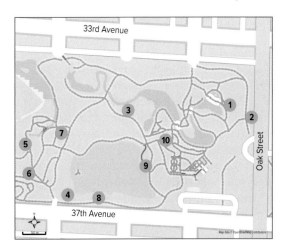

1. *Gymnocladus dioica*—Kentucky coffeetree
2. *Castanea dentata*—American chestnut
3. *Picea chihuahuana*—Chihuahuan spruce
4. *Styphnolobium japonicum*—scholar tree
5. *Zanthoxylum schinifolium*—Szechuan pepper
6. *Broussonetia papyrifera*—paper mulberry
7. *Taiwania cryptomerioides*—coffin tree
8. *Carrierea calycina*—goat horn tree
9. *Fagus sylvatica* 'Bornyensis'—Borne weeping beech
10. *Pinus wallichiana*—Himalayan pine

Queen Elizabeth Park
4600 Cambie St

The highest point in the city, called Little Mountain by some, this park includes an arboretum first planted in 1949 with trees from around Canada. The concept has loosened since then to include exotic trees from everywhere, alas not many with labels. This tour offers a return to the original idea in narrower form by mapping a selection of trees native to this area. A good start for native plant enthusiasts and anyone else curious to know what this bioregion is all about.

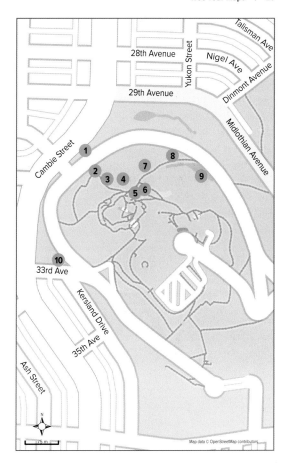

1. *Picea sitchensis*—Sitka spruce
2. *Acer circinatum*—vine maple
3. *Pseudotsuga menziesii*—Douglas-fir
4. *Acer macrophyllum*—bigleaf maple
5. *Thuja plicata*—western redcedar
6. *Malus fusca*—Pacific crabapple
7. *Cornus nuttallii*—Pacific dogwood
8. *Tsuga heterophylla*—western hemlock
9. *Populus trichocarpa*—black cottonwood
10. *Chamaecyparis nootkatensis*—yellow-cedar

The Crescent—Shaughnessy Park
1300 The Crescent @ Hudson St

A pleasantly treed park in Shaughnessy, a neighbourhood developed and marketed by the Canadian Pacific Railway in 1907. They were on to a good idea when they designed this oval as a green centrepiece. Many of the trees in the shady space (and along the streets leading into it) have grown to an impressive size. Although it may be true that west coast forests tend to pale compared to the brilliant autumn colours of those back east, the spreading sugar maples and sourwoods planted here put on a splendid display in good years.

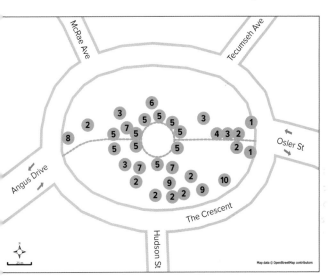

1. *Acer saccharum*—sugar maple
2. *Ulmus carpinifolia*—smoothleaf elm
3. *Picea abies*—Norway spruce
4. *Chamaecyparis pisifera*—Sawara cypress
5. *Tilia platyphyllos*—largeleaf linden
6. *Pseudotsuga menziesii*—Douglas-fir
7. *Oxydendrum arboreum*—sourwood
8. *Sequoia sempervirens*—redwood
9. *Cladrastis kentukea*—yellowwood
10. *Fagus sylvatica*—European beech

Kitsilano Beach
Cornwall Ave @ the N end of Yew St

The beach gets most of the attention here from the droves who visit, and not only in summer, but the trees help make the site impressive. They frame some beautiful seascapes or, if you're paying attention, can be a view of beauty on their own. There are some classic elms here, the always dramatic willows and a few supermodels of the flowering tree world that can stop traffic when in bloom—check out the Akebono cherry and the prostrate yet graceful purpleleaf plum (not a cherry despite what most people think when they see it bloom). The first row of trees leading into the park from Yew St are horsechestnuts which, alas, didn't make the list.

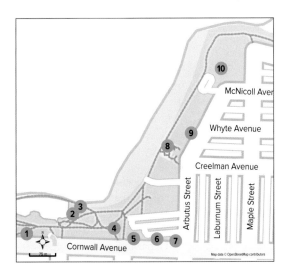

1. *Prunus cerasifera* 'Atropurpurea'—purpleleaf plum
2. *Magnolia grandiflora*—evergreen magnolia
3. *Ulmus x hollandica*—Dutch elm
4. *Pterocarya stenoptera*—Chinese wingnut
5. *Prunus yedoensis* 'Akebono'—Akebono cherry
6. *Fraxinus ornus*—flowering ash
7. *Ulmus carpinifolia*—smoothleaf elm
8. *Salix x sepulcralis* 'Chrysocoma'—golden weeping willow
9. *Robinia pseudoacacia*—black locust
10. *Castanea sativa*—sweet chestnut

Treasured Trees

Vancouver has about 500,000 trees and 500 tree species (we won't even guess the number of cultivars). To think that 10 trees could represent a forest so grand would be daft. Rather let's consider this a snapshot of trees worth seeing, and a base for you to create a map of your own worth cherishing and sharing.

1. *Thuja plicata*—western redcedar. The youngest tree in this book, planted in Oct. 2015 to honour environmentalist David Suzuki—N edge of the City Hall grounds, N of the demonstration food garden.

2. *Quercus robur*—English oak. Planted for King George VI on coronation day May 12, 1937—NE corner of Cambie St and 12th Ave on City Hall grounds.

3. *Pterocarya fraxinifolia*—Caucasian wingnut. A fine tree even without the delightful common name, this one was said by a local, who heard it from an elderly neighbour, to have turned 100 in 2010—SW corner of Chilco St and Comox St.

4. *Quercus rubra*—red oak. The first Vancouver PoetTree inaugurated in 2013 (see Trout Lake map for the present PoetTree)—Bidwell St and Beach Ave in SW corner of Alexandra Park.

5. *Sorbus x thuringiaca* 'Fastigiata'—oakleaf mountain ash. Centre of the Medicine Wheel Healing Garden in Oppenheimer Park in the NE area of the park E of the fieldhouse.

6. *Fagus sylvatica* 'Asplenifolia'—fernleaf beech. On the NW side of Grandview Park above the stairs to Cotton Dr.

7. *Cedrus deodara*—deodar cedar. Near the SE corner of Victoria Dr and 41st Ave in the parking lot, showing how a

lone specimen can grace a drab urban intersection with the beauty of the Himalayas.

8. *Prunus yedoensis 'Akebono'*—Akebono cherry. Queen Elizabeth Park S of Bloedel Conservatory, a tree so attractive in bloom you have to wait in line to take its picture.

9. *Liriodendron tulipifera*—tuliptree. In the yard at 6964 Yew St across from the SE corner of Maple Grove Park at SW Marine Dr and Yew St.

10. *Abies grandis*—grand fir. In the old growth forest at the W edge of UBC. From the intersection of SW Marine Dr and the W end of Old Marine Dr (N of the botanical garden), look into the ravine to see the oddly erratic crown of perhaps the third tallest grand fir in the province. Big tree researcher Ira Sutherland says the scraggly leaders on the "Totem Giant," like those on other grand firs in the area, are due to an attack of balsam wooly adelgid. Click vancouverbigtrees.com for more on the trees here. Visitors in warm weather should know Wreck Beach is clothing optional.

Old growth *Pseudotsuga menziesii*—Douglas-fir (No. 4 in Stanley Park Tree Tour).

Tips to ID Trees

- Use all your senses—touch the bark, smell the leaves, taste the fruit (non-poisonous only!).
- Consider a small hand lens (10×) to check details, binoculars to see canopy flowers or cones.
- Bring a bag or container to take leaf/flower/cone specimens home for further study.
- Don't fret over a hard-to-figure type…even tree experts get stumped.

1 *Araucaria araucana*
monkey puzzle tree

LEAVES Evergreen with big overlapping scales too sharp to grab.

FORM Tall and imposing with a single, straight trunk holding long, horizontal, whorled branches; crown spreads with age.

BARK Sharply spiked when young, then grey and fissured.

FLOWERS/FRUIT Male cones oblong, 8–12 cm, upright, brown, scaly; female cones the size of a coconut; seeds big and edible.

Native to Chile and western Argentina but endangered there by illegal logging and shrinking forest spaces. A fad for this unusual tree swept 19th century England where its common name was suggested for spiny foliage that might repel even a monkey. Some Vancouverites planted it in their yards in the early 20th century so we have a few sizeable examples of this tall, twisted wonder, although nothing like the 40-m giants of its homeland. The seeds (from female trees only) were a mainstay of the pre-European Pehuenche ("People of the Trees"). Bigger than almonds, they're tasty raw, and can be cooked or fermented into a traditional Pehuenche drink called *mudai*.

NE corner of W 21st Ave and Columbia Ave • Burnaby St W of Bute St.

LEAVES Evergreen scales in flattened sprays; bright green.

FORM Stiff, upright column or narrow pyramid to 30 m tall; gracefully drooping branches.

BARK Reddish-brown, ridged, shredding.

FLOWERS/FRUIT Small yellow male cones on branch tips in winter; seed cones 2–3 cm long, narrow, with six scales (fewer than cones of similar-looking western redcedar) which open to resemble a woody duck's bill.

Native to mountain areas of California and Oregon. Earns the name for the arresting aroma from leaves, sap and wood, which might invoke memories of school if you used No. 2 pencils.

In the lawn across the road from the Stanley Park Pitch 'n' Putt front gate is a tall incense cedar next to a giant sequoia and a redwood.

Chamaecyparis lawsoniana
Lawson cypress

LEAVES Evergreen fan-like sprays of fine, overlapping scales; blue-green or green, whitish beneath; resiny aroma.

FORM Narrow conical crown to 60 m or taller with branches typically reaching to the ground; trunks often buttressed.

BARK Thin, reddish-brown when young; greying and ridged with age; peeling in flat, vertical strips.

FLOWERS/FRUIT Male pollen cones tiny, reddish, seen in winter at branch tips; female seed cones blue-green, maturing to little brown soccer balls less than 1 cm diameter.

Native to a few areas of the Pacific coast from Southern Oregon to Northern California. Not a huge natural habitat, and the trees are in trouble in the wild as old growth forests have been clear-cut. Even trees planted in gardens on native root stock are threatened by a fatal root disease known as *Phytophthera*. On the bright side, Lawson cypress have become so popular among gardeners that they're now planted widely in resistant cultivars of all shapes and colours. Larger specimens make good wildlife trees: raccoons turn them into housing complexes where they can sleep undisturbed all day even in busy urban areas. Also called Port Orford cedar, white cedar.

2056 Grant St W of Lakewood Dr • W 16th Ave W of Maple St.

1

Chamaecyparis nootkatensis
yellow-cedar

LEAVES Evergreen scales in alternating pairs of four even rows along the rounded twig (compare to the flatter twigs of western redcedar with two rows of flattened scales and two rows of folded scales); rough texture if you rub back "against the grain" (redcedar is flat and smooth); green to yellow-green; aroma not very pleasant, unlike redcedar.

FORM Narrow cone to 25 m tall with swooping downward branches that curve gracefully back up to hold drooping branchlets.

BARK Greyish-brown, irregularly fissured, stringy.

FLOWERS/FRUIT Seed cones look like small green berries to start, maturing the next year to woody, brownish balls with 4–6 scales (compare to western redcedar which have oblong cones).

Native from Alaska to Oregon including the mountains around Vancouver. Yellow-cedar is planted in gardens for its graceful beauty, usually in cultivar form, of which there are many types. Favoured in indigenous circles for long, straight, durable wood that can be carved into paddles, masks, totem poles, boxes and much more. Leaves may resemble those of western redcedar, but note that they have rough twigs (not soft) and the tree's bark doesn't peel (redcedar bark comes off in strips). A venerable native tree that may live thousands of years, and then lie on the forest floor for another millennium, like redcedar, thanks to rot-resistant wood. Also called Nootka cypress, yellow cypress, Alaska cedar.

1915 Haro St W of Gilford St has a short linear grove • the even-droopier Pendula cultivar (weeping yellow-cedar) can be seen in the NW corner of McCleery Park at the corner of W 49th Ave and Marine Cres.

1

Chamaecyparis obtusa

hinoki cypress

LEAVES Evergreen scales in flattish, fan-like sprays; dark green with white lines beneath; tips are blunt or "obtuse" as in the name.

FORM Conical, to 35 m high.

BARK Dark, reddish-brown.

FLOWERS/FRUIT Seed cones medium-sized brown globes 8–12 mm across, with 8–12 scales.

A graceful tree revered in Japan for its fragrant wood and used to make temples, shrines and traditional baths. Hinoki groves in Japan are prime sites for the practice of *shinrin yoku*, or "forest bathing." It describes walking in the forest to improve one's mind/spirit/body balance. People have done this intuitively for thousands of years but scientists are now exploring how "terpenes," chemicals released from the trees, have a measurable affect on human health. The Japanese government has recognized 60 Forest Therapy Bases nationwide. Hinoki here are seen more often in cultivar form, including the shrubby Compacta and its golden cousin Lutea.

 2014 E Pender St in the median.

LEAVES Evergreen overlapping triangular scales curved like awls; longer on new shoots; blue-green or green.

FORM Massive conical shape with horizontal (and easy to climb) branches reaching to the ground; rounder and less regular with age.

BARK Reddish-brown maturing to grey, deeply fissured; very thick in older trees.

FLOWERS/FRUIT Egg-shaped cones 5–10 cm long at the ends of shorter shoots.

Although not the tallest tree in the world—that's the redwood—giant sequoia is the largest in volume. The trunk alone can be wider than a tennis court and tree heights top out above 90 m. Native to the western slopes of the Sierra Nevadas in California, it thrives when introduced to congenial places such as here. Like a whale, a single specimen can astound with size alone, but the more you look, the more you see the grace beyond sheer mass. Giant sequoia live for centuries, thanks in part to a thick, spongy bark that can endure most forest fires. Weeping giant sequoia, the Pendulum cultivar, is not a giant at all but a bendy stick with needles discovered as a mutant in France and now planted widely for quirky effect, like something Dr. Seuss brought to life.

2307 W 41st Ave • a grove at VanDusen Botanical Garden shows how much they've grown since a 1973 planting—and these mammoths are still adolescents.

1 *Thuja plicata*
western redcedar

LEAVES Evergreen with flattened, droopy, fern-like sprays of yellow-green scales that alternate in overlapping pairs of flat and folded rows; aromatic when crushed.

FORM Grows to majestic proportions with spreading branches turning up at the tips; often with a buttressed trunk holding a long, symmetrical, narrowly conical crown; grows less regular with age.

BARK Smooth, shiny brown when young; matures to a fibrous, peeling reddish-brown that gleams in a sunbeam.

FLOWERS/FRUIT Small 1-cm-long egg-shaped cones with 8–12 scales, often with a sharp point near the tip, appearing in clusters; maturing to a woody brown and turning upward.

If you learn to recognize just one Vancouver native tree, make it western redcedar. In turn it could become your favourite tree, and if you live here, one that will always remind you of home. Vital to First Nations culture for thousands of years, western redcedar is a cornerstone species in indigenous society. The trees provide soft yet durable wood for homes, canoes, paddles, totem poles and much more. The bark can be woven into rain-repelling clothes and hats, while the inner bark and needles

E side of NW Marine Dr where it bends at Locarno Beach • throughout wooded areas of Stanley Park, Pacific Regional Spirit Park, Queen Elizabeth Park and citywide.

provide medicine and food. Western redcedar is the official tree of British Columbia, and rightly so; no other tree speaks more eloquently of the rainforest in this part of the world.

Sequoiadendron giganteum—giant sequoia

2 *Abies amabilis*
amabilis fir

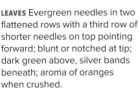

LEAVES Evergreen needles in two flattened rows with a third row of shorter needles on top pointing forward; blunt or notched at tip; dark green above, silver bands beneath; aroma of oranges when crushed.

FORM Up to 50 m tall, dense and narrow symmetrical cone with tiered branches.

BARK Smooth, grey-white, scaly with age.

FLOWERS/FRUIT Purple, barrel-shaped seed cones held upright on branches (typical of firs) at tops of trees; 9–14 cm long.

Native to higher elevations in our area, they're more in their element in the North Shore mountains than in the city lowlands. Indigenous uses include spreading the

boughs for aromatic flooring and bedding—try it at home. Also used to treat colds and flu with a tea steeped from the needles. The botanical name *amabilis* means lovely, a tribute to its pleasing shape. Also known as Pacific silver fir for the silvery lines of stomata on the leaf undersides.

 Two on N side of 38th Ave between Dunbar St and Collingwood St.

LEAVES Evergreen needles flat, soft, 4–7 cm long (longer than most firs), bluish-green.

FORM Medium to tall (to 30 m in the wild), symmetrical, branched to near ground level; mature trees get flatter tops and drooping branches.

BARK Rough, light grey, thick ridges with age.

FLOWERS/FRUIT Greenish cylinders, turning brown with age, 7–12 cm long; held upright on branches on top of the tree.

Native to the Sierras and southern Rocky Mountains, and planted widely in gardens elsewhere for their beauty in a variety of cultivars. Unlike most firs, which have needles that are dark above and whiter beneath, these are the same hue (concolor) on both sides. Fir trees are typically happier in the mountains than the city but this type may be attractive enough to try if you have a site with well-draining soil. The noble aspect and great smell of fir trees make them popular house guests at Christmas.

 SW corner of 3rd Ave and Waterloo St amid other tall trees.

2 *Abies grandis*
grand fir

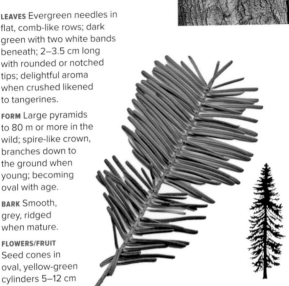

LEAVES Evergreen needles in flat, comb-like rows; dark green with two white bands beneath; 2–3.5 cm long with rounded or notched tips; delightful aroma when crushed likened to tangerines.

FORM Large pyramids to 80 m or more in the wild; spire-like crown, branches down to the ground when young; becoming oval with age.

BARK Smooth, grey, ridged when mature.

FLOWERS/FRUIT Seed cones in oval, yellow-green cylinders 5–12 cm long with scales wider than long; held upright on branches high up on the tree.

The common name is apt—majestic in size and shape, this tree is indeed grand and one of our most attractive native species. Used by First Nations for dyes, tonics, fish hooks, paddles and canoes. The aromatic branches can also be woven into ceremonial costumes and rubbed onto the skin before rituals. Blisters in the bark of new growth can be popped to apply the therapeutic resin onto insect bites and small cuts.

A row at 4796 W 4th Ave • UBC has a goliath on West Mall south of the intersection at Memorial Rd, and more in the old growth forest in the ravine off Wreck Beach.

Tips to Identify Fir Trees

- dense, compact, spire-like crowns
- upright cones
- cones that disintegrate at maturity, leaving a single spike behind
- needles that are flat, often with a blunt or notched tip
- needles that attach directly to twig, leaving a round, depressed circle when removed
- needles that may show white lines of stomata (pores)
- needles that are soft, attractive, aromatic

2 *Cryptomeria japonica*
Japanese cedar

LEAVES Evergreen pointy, curved needles, arranged in overlapping spirals; 0.5–1 cm long.

FORM Up to 70 m in a tall, narrow cone; trunk up to 4 m in diameter.

BARK Reddish-brown, fibrous.

FLOWERS/FRUIT Seed cones 1–2 cm long, green turning to brown; pollen cones 0.5 cm long, purple-brown, clustered on the tips of two-year-old branches.

Native to Asia. In Japan it's the national tree called *sugi* and much admired for its fragrant, resilient wood. A feudal lord honoured the death of the Tokugawa shogun in 1616 by planting an allee of Japanese cedar. It's still there today, the longest tree-lined avenue in the world, according to the Guinness book. Start planting now if you hope to challenge it—the record is 35 km long with 13,000 trees. Hundreds of cultivars of varying shapes and colours have been developed, although the original is still hard to beat for beauty.

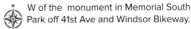

W of the monument in Memorial South Park off 41st Ave and Windsor Bikeway.

LEAVES Evergreen with small, sharp-tipped needles arranged in whorls of three; fragrant.

FORM Usually a low, spreading shrub but occasionally grown as a small tree.

BARK Brown and fibrous, exfoliates in strips or plates.

FLOWERS/FRUIT Fleshy seed cones (technically not "berries") go from bluish-green to blackish-purple.

Common juniper is the most common and widespread conifer in the world. Native to Europe, Asia and North America including BC, the juniper genus has some 60 species and enough cultivars to confuse the most determined taxonomist. Native uses for common juniper include treating cuts and burns with the sap, using the inner bark for stomach troubles and rubbing a concoction from the steamed roots into the skin for arthritis. Drinkers will appreciate knowing the seed cones from common juniper give gin its flavour.

Although not popular today in garden stores, many were planted in previous decades and can be seen in private gardens • VanDusen Botanical Garden has some that show the typical low-growing aspect, and a taller Chinese juniper (*Juniperus chinensis* 'Keteleeri') can be seen NE of the Rose Garden in Queen Elizabeth Park.

2 *Metasequoia glyptostroboides*
dawn redwood

LEAVES Deciduous soft, flat needles to 2.5 cm long arranged in opposite rows along the twig; emerging light brown in spring, turning to green in summer and then brown before dropping in fall.

FORM Tall, conical pyramids for several decades, becoming less regular with age; to 30 m or more.

BARK Reddish-brown, fibrous.

FLOWERS/FRUIT Solitary oval cones 2–3 cm long on 5-cm-long stems.

Fossils showed that dawn redwood once spread around the world, including here in Vancouver, but it was thought to have gone the way of the dinosaur until 1941 when a few living specimens were found in a remote valley in China. The news, and seeds, spread widely, and we can now enjoy these graceful, deciduous conifers even here. Worth planting if you have room for a gorgeous but big tree. The new needles in early summer make a tasty addition to salads.

 Kerr St between 41st Ave and 43rd Ave has a lovely and rare example of a street filled with dawn redwoods.

Metasequoia glyptostroboides—dawn redwood

2 *Picea abies*
Norway spruce

LEAVES Evergreen needles stiff, sharp, 1–2.5 cm long; dark green.

FORM Pyramidal crown to 40 m tall; ascending main branches with drooping secondary branches.

BARK Reddish-brown to darker grey when mature, smooth turning to scaly.

FLOWERS/FRUIT Large cylindrical cones 10–15 cm long (longest of the spruces); scales pointed; light green turning to brown.

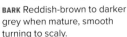

The tallest native tree in Europe is commonly planted here, in part because it's tough—our native Sitka spruce does not adjust as well to lowland city life. Hundreds of types, including some dwarf cultivars which may make a more reasonable fit for a residential yard. Spruce are important medicinal trees in traditional indigenous healing. The sap can be used to treat sores, inflammation, blood poisoning, arthritic joints and heart trouble. Early European visitors warded off scurvy with spruce beer; some claim it tastes like root beer. Spruce gum made from the pitch can be chewed as a sugary-gum substitute or boiled into cough syrup to relieve colds and sore throats.

5782 Ontario St at Ontario Pl • many other sites citywide including near the gate of the Stanley Park Pitch 'n' Putt just south of a large *Sequoiadendron*.

2 *Picea glauca*
white spruce

LEAVES Evergreen needles arranged spirally on short stalks; four-sided in cross-section; 1.5–2 cm long; bluish to dark green with white dots.

FORM Medium to large tree up to 40 m with a narrow, conical crown; branches slightly drooping.

BARK Smooth, light grey when young, darker and scaly with age; exposed bark silvery to orangish-pink.

FLOWERS/FRUIT Slender, cylindrical, 3–6 cm long; purple or green maturing to reddish-brown.

A hardy species that ranges up to the arctic tree line, white spruce is native to forests throughout Canada except for the west coast, and is an important commercial tree for lumber and pulp. Not planted often but the dwarf cultivar Conica is popular as a tidy yard evergreen. Hybridizes naturally with other spruce trees, leading to variations that are hard to figure out. Not as prickly as most spruce needles. Some find the needles smell less than pleasant when crushed; it may depend on whether you like camphor. Also called Alberta white spruce, western white spruce, Canadian spruce and skunk spruce.

A few can be found in the VanDusen Botanical Garden's conifer collection, although you need to look for them.

2 *Picea omorika*
Serbian spruce

LEAVES Evergreen needles flat (not four-sized like most spruces) and flexible with tips that are blunt (not sharp like most spruces); 1–2 cm long; bluish-green.

FORM Narrow pyramid with short, drooping side branches that curl up at the tips; medium-sized to 30 m tall.

BARK Thin, brown, scaly plates.

FLOWERS/FRUIT Seed cones oblong, to 6 cm long and pendulous, turning from purple to reddish-brown.

A fine choice for year-round greenery in a tight urban space. The tall, tight and graceful form makes it a favourite of conifer aficionados. Native to Bosnia and Serbia, it was introduced to the horticulture trade in the 1880s. It spread widely, which was fortunate because most have died in its native habitat from development and fires including those sparked by the war in Bosnia.

Grandview Highway median just W of Nanaimo St • in front of the apartment building at 1556 Charles St.

Tips to Identify Spruce Trees

- needles stiff and sharp
- needles 4-sided and can be rolled between the fingers
- needles held singly on peg-like bases
- bare twigs stiff and stubby with rasp-like pegs
- cones hang down and have thin, papery scales
- bark typically thin and scaly

2 *Picea pungens*
blue spruce

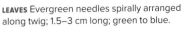

LEAVES Evergreen needles spirally arranged along twig; 1.5–3 cm long; green to blue.

FORM Dense pyramid with horizontal lower branches; to 30 m tall.

BARK Grey to brown, flaky, furrowed with age.

FLOWERS/FRUIT Cylinder-shaped cones with tapered tips, 5–10 cm long; scales thin, flexible, loose.

Native to the Rockies from Colorado to Utah but widely planted elsewhere (some say over-planted), especially the varieties with eye-catching silvery-blue needles, which are what most people want. The specific name *pungens* refers to the sharpness of the needle tips. Also called Colorado spruce.

NW corner of Victoria Dr and Grant St • Tisdall St N of 50th Ave.

LEAVES Evergreen needles arranged spirally along twig; flattened (unlike most spruces), stiff with a sharp tip; 2–3 cm long; dull green above, white stomata lines beneath.

FORM Robust pyramid up to 90 m or taller with massive trunk, often buttressed; somewhat open crown.

BARK Thin, loose, plate-like scales, grey to reddish-brown.

FLOWERS/FRUIT Reddish- to yellowish-brown cylindrical seed cones 5–10 cm long; seed scales thin and wavy with irregularly toothed edges.

Native to the Pacific coast, this is the largest of the spruces. The Carmanah Giant on Vancouver Island is monumental, perhaps the tallest tree in Canada at 95 m (there may be taller ones either undiscovered or unannounced by wary tree hunters). Indigenous uses for Sitka spruce include making baskets and hats from the roots and branches, and turning the sap into medicine. The inner bark in spring can be eaten fresh or dried into cakes, and steeped needles make a tasty tea. A valuable commercial tree not only for volume; the light yet strong wood was once used to build airplanes.

Several at Dunbar St and W 51st Ave beside the golf course • in Queen Elizabeth Park W of the park roadway about 40 m W of the Cambie St entrance.

2 *Pseudotsuga menziesii*
Douglas-fir

LEAVES Evergreen needles soft, flat, arranged spirally along the branch like a bottlebrush; 2–3 cm long with prominent mid-vein, narrowing base and pointed tip; yellow-green above and paler beneath with two lines of white dots; pleasant aroma when crushed.

FORM Narrow conical shape when young, maturing into a pyramid atop a massive bare trunk; may become flattened on top.

BARK Smooth, grey-brown, resin-blistered when young; thick and deeply furrowed when mature, the corky cover helps the tree survive forest fires.

FLOWERS/FRUIT Woody, 6–9 cm long, reddish-brown cones hang down (true firs bear upright cones). If none are in reach to check, search the base of the tree for fallen cones; the three-pronged bracts are a good ID clue if you remember the quaint tale that deer mice, friends of Douglas-fir, hide from foxes by partially burrowing into their cones.

A native giant that helps define the Pacific Northwest. Not a true fir (hence the hyphenated common name) and not a hemlock either (the genus name *Pseudotsuga* means "false hemlock"). The species name *menziesii* honours Scottish surgeon-botanist Archibald Menzies who sailed with George Vancouver here in 1792, and was once locked in the brig by the hotheaded captain for complaining about the mistreatment of his collected plants. Indigenous uses for Douglas-fir include making spears, bowls, snowshoes and canoes. In 1978 Geordie Tocher and two others left Vancouver in a Douglas-fir dugout canoe named Orenda II, reaching Hawaii 54

days later. Dried sap can be chewed to relieve symptoms of a cold and the sticky buds used to heal mouth sores. The aromatic foliage makes good bedding or can be rubbed onto the skin to prevent deer from picking up the scent of human hunters. Some Douglas-fir trees have survived more than 1,000 years, overcoming all challenges but dunderheads. The Stanley Park Ecological Society says a venerable 99-m-tall Douglas-fir that used to tower over the entrance to the park was cut down for firewood. Many remain in the park, including some old-growth giants that somehow escaped the axe.

In forested areas citywide • a little-known old growth grove in the cliffs off Wreck Beach at UBC includes some Douglas-fir trees more than 400 years old

2 *Sciadopitys verticillata*
umbrella pine

LEAVES Evergreen needles in umbrella-like whorls; flat, 8–16 cm long; bright green above, paler beneath with white stomata lines.

FORM Up to 30 m; small- to medium-sized conical shape with short, dense, horizontal branches.

BARK Reddish-brown, fibrous.

FLOWERS/FRUIT Seed cones oval, 8–12 cm long, scales fan-shaped and flexible.

A true beauty from Japan, not widely sold here and often expensive when it can be found because a very slow growth rate makes it unappealing to nursery suppliers. Not actually a pine, it's classified in its own family, Sciadopityaceae, and considered a living fossil with records showing it spread around the earth even before dinosaurs. Its natural range today is limited. In Japan it is known as *koya-maki*, and considered sacred by some. A single Japanese umbrella pine has been worshipped at a Kyoto temple since 1310. It has a reputation for being a small tree here but Portland south of the border has examples of umbrella pines towering over the houses of their owners.

The yard at 3669 W 1st Ave has one registered as a Vancouver Heritage Tree, a list of worthy trees compiled in 1983 as a project for the BC Heritage Trust.

2 *Sequoia sempervirens*
redwood

LEAVES Evergreen short, slender, pointed, green needles with white stomata dots beneath; new side shoots in flat sprays with scale-like leaves.

FORM Tall, elegant, pyramid with some drooping main branches when young; maturing to a more rounded top and fewer lower branches.

BARK Reddish-brown, fibrous, very thick.

FLOWERS/FRUIT Pollen cones small, yellow, clustered at branch tips; seed cones hang on the ends of shorter shoots, maturing from green to brown in one season.

The world's tallest tree, Hyperion in northern California, wins the height contest at 115 m (the Leaning Tower of Pisa is 55 m). The biggest redwoods were almost completely logged out in their native habitat in the fog belt of coastal northern California and Oregon. Stories persist of felled trees much larger than the few titans left today. Introduced to cultivation in 1843 and now widely planted, including in Vancouver. Maybe in a few centuries we'll have some cloud-breakers here as well. Also called coast redwood, common redwood and California redwood.

 SW corner of W 13th Ave and Pine St • Thornton Park at Main St and National Blvd.

2 *Taxus baccata*
English yew

LEAVES Evergreen needles similar to the native Pacific yew with dark green and glossy surfaces, 1.5–2 cm long, tapering to a point.

FORM A low shrub or obedient hedge when pruned, but can grow to 10 m if left alone.

BARK Brownish-red, furrowed and flaky; less colourful than the native Pacific yew.

FLOWERS/FRUIT Soft coral-coloured cup known as an aril holding a single (poisonous) seed.

English yew have a long and storied history in Europe for the sturdiness of their wood which was ideal for bows and other weapons. Less know is that all parts of the plant are poisonous. The flesh of the red berry-like aril may not actually kill you—the poison is said to be in the hard seed that should pass through you undigested—but cows have died from nibbling on the leaves. Difficult to distinguish from the Pacific yew with both types making their way in the wild and probably hybridizing, but most

Oak St N of Broadway at BCAA parking lot entrance • Queen Elizabeth Park at Cambie St and W 33rd Ave, behind benches in the Quarry Gardens.

of the hedges and garden fixtures seen here are cultivars of English yew. Yew trees have been known to live for more than 1,000 years. Ancient Celts revered them, and were known to touch trees in times of need. A remnant of this practice may be reflected in our good-luck trick to "knock on wood."

2 *Taxus brevifolia*

Pacific yew

LEAVES Evergreen needles arranged spirally but flattened on either side of the twig so they appear to be in rows; 1.5–2 cm long with pointed tips; yellow-green above, paler with whitish bands below.

FORM Small tree with a twisting trunk and horizontal branches to 20 m, or a low shrub.

BARK Smooth, thin, reddish-brown when young, gets shaggy with age and exfoliates to reveal rosy bark beneath.

FLOWERS/FRUIT Soft coral-coloured cup known as an aril holding a single (poisonous) seed.

Pacific yew are understory trees found from Alaska down the coast to central California and in the Rocky Mountains south to Idaho. The wood is dense and hard, preferred by First Nations for tools that must perform under pressure such as bows, clubs, paddles and digging sticks. Used in traditional medicine for centuries, western science more recently caught on by developing an extract from the bark called taxol as an anti-cancer agent. It became the best-selling cancer drug ever, and is now available in synthetic form.

See the Canadian Heritage Garden at VanDusen Botanical Garden.

2 *Tsuga heterophylla*
western hemlock

LEAVES Evergreen needles flat and soft with blunt tips, arranged in opposite rows (unlike mountain hemlock which cover the twig all around); shiny green on top, pale white dots below; unequal lengths (hence the name heterophylla) from 0.5–2 cm.

FORM Large, to 50 m tall, and graceful with a long bare trunk leading to an open, conical crown, often with a tell-tale drooping tip; swooping feathery branchlets add to the languid appeal.

BARK Smooth, reddish-brown when young, becoming darker and furrowed with age.

FLOWERS/FRUIT 1.5–2.5 cm oval, brown seed cones on short pegs.

Able to grow in shade, a good strategy to succeed in a dense rainforest, they're often seen as young seedlings in the tops of logged stumps because that's where they find moisture. Hemlocks are like standing medicine chests: the bark can treat sores, broken limbs, heart trouble and more, while the inner bark (cambium) has been baked or steamed into a nutritious staple food for many generations. The needles, high in vitamin C, may be brewed into a rejuvenating tea; hikers have been known to chew the tips of new leaves to suppress hunger. An all-purpose resilient tree good for everything except firewood; one First Nation name for it translates to the "no fire tree."

NE corner of Cotton Dr and William St • throughout Stanley Park and other wooded sites, which would naturally be hemlock forest if there were no interventions.

Tips to Identify Hemlock

- drooping tip
- soft, lacy foliage
- needles with short stems held on woody cushions
- cones hang from tips of branches

2 *Tsuga mertensiana*
mountain hemlock

LEAVES Evergreen needles crowded and spreading around the twig like a bottlebrush; 2–3 cm long with blunt tips; dark green with faint white dots above and below.

FORM Small- to medium-sized tree that can grow to 40 m with slightly drooping tip and branches that turn up at the tips; narrow conical crown gets irregular with age.

BARK Dark reddish-brown, scaly.

FLOWERS/FRUIT Cylindrical purplish-brown seed cones 3–7 cm long (longer than western hemlock); fan-shaped scales.

A native beauty but seen less often in Vancouver than the surrounding mountains where it grows best. Needles are on small stalks, like spruce, but unlike spruce are not prickly. Unlike western hemlock needles, which have stomata on the underside of needles only, mountain hemlock show faint white dots on both upper and lower sides. For further data, cut a needle in two to see how it's round in cross-section (western hemlock needles are flatter).

2893 W 20th at MacKenzie St.

3 *Cedrus atlantica*
Atlas cedar

LEAVES Evergreen needles in clusters; less than 2.5 cm long, dark green or, in the popular Glauca variety, ethereal silvery blue.

FORM Large and stately pyramid 15–40 m tall, growing less regular with age.

BARK Grey, fissured, breaks into scaly plates.

FLOWERS/FRUIT Cones big and barrel-shaped to 7.5 cm long and held upright; green turning to grey; crumbling on the branch rather than falling intact, leaving a spike.

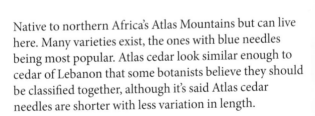

Native to northern Africa's Atlas Mountains but can live here. Many varieties exist, the ones with blue needles being most popular. Atlas cedar look similar enough to cedar of Lebanon that some botanists believe they should be classified together, although it's said Atlas cedar needles are shorter with less variation in length.

NW corner of Kingsway and Moss St • for *Cedrus atlantica* var. *glauca* (blue atlas cedar) see SE corner of Granville St and 49th Ave or the one espaliered on the front wall of the Vancouver Park Board admin office at 2099 Beach Ave.

3

Cedrus deodara
deodar cedar

LEAVES Evergreen needles varied in length 2.5–5 cm and emerging in clusters, shorter towards the branch tips; blue-green.

FORM Large pyramids to 25 m or more; gracefully tiered branches with drooping tips; drooping leader as with hemlocks.

BARK Smooth, grey when young, maturing to furrowed with scaly ridge tops.

FLOWERS/FRUIT Large pale green cones to 10 cm turning reddish brown after two years.

Also called Himalayan cedar, this striking tree evokes the majesty of that mountain range wherever it gets planted…so long as it has room. The long sweeping branches may continue down to ground level, making this a tree to stand back and admire. From the Sanskrit for "tree of the gods." May be confused with other cedars introduced here but deodars have longer needles and bigger seed cones than the rest.

Cambie Heritage Blvd median N of Kingsway

3 *Cedrus libani*
Lebanon cedar

LEAVES Evergreen needles in clusters in varying lengths 2–3 cm long; bright green when new turning to dull green.

FORM Strong single or multiple trunks with wide, spreading, horizontal branches.

BARK Grey, fissured, breaks into scaly plates.

FLOWERS/FRUIT Egg-shaped cones are similar to but larger than Atlas cedar and smaller than deodar cedar.

True cedar trees come in just four species, although the common name is used often for other trees that happen to be beautiful or fragrant. Cedars have a long history in the Old World as sacred. Greek writer Philostratus describes a rich man in what is now Turkey who lost a good portion of his wealth in the year 200 through fines for cutting down cedars. King Solomon is said to have used this tree to build his temple in Jerusalem. You can see it on the flag of Lebanon, the only country to so honour a complete tree (Canada gets a nod for depicting a leaf). There aren't many in Vancouver as it doesn't always do well here, but well-tended trees in Queen Elizabeth Park and VanDusen Botanical Garden are worth visiting.

Queen Elizabeth Park south of quarry, with a plaque explaining the importance of this species to sea-faring explorers.

3

Larix occidentalis
western larch

LEAVES Deciduous needles in tufts of 15–30; triangular in cross-section; 3–5 cm long.

FORM Up to 70 m tall with a long branch-free trunk leading to a short pyramidal crown.

BARK Reddish-brown, deeply furrowed with age.

FLOWERS/FRUIT Cones 3–5 cm long with a long bract tip extending beyond the scales.

Native to the BC interior, but not here. Larch trees look like an evergreen until they naturally drop their needles each fall, shocking some who think the tree just died. Even arborists have been known to remove a "dead" larch in winter due to mistaken identification. Cousins include European larch (*Larix decidua*), which can be seen in a grove in Queen Elizabeth Park, and hybrid larch (*Larix x eurolepis*), with a few beside the parking lot in VanDusen Botanical Garden. The bright and shiny new needles are a visual treat in spring, making the barren months seem worth the wait.

East side of Elliott St S of 54th Ave, along with some *Larix Kaempferi* (Japanese larch) which have a bluer tint to the needles, a pinker tint to the twigs and stubbier cones with scales that curve outward.

3 *Pinus contorta* var *contorta*
shore pine

LEAVES Evergreen needles in bundles of two; sharply pointed, 3–5 cm long; often twisted.

FORM Shrubby or small tree to 10 m tall; branching may look scraggly and windswept.

BARK Reddish-brown, thick with deep grooves.

FRUIT/FLOWERS Cones oval 3–5 cm long; prickly point at the tip of each scale.

The only pine native to the city area. On a windswept beach it can look haggard, as if designed to hunker through a gale. When grown close together, the variety known as lodgepole pine produces tall, straight trunks worthy of the name; teepee builders would use them in home construction. Indigenous uses for shore pine include medicines derived from the bark and sap. You can chew the buds to relieve a sore throat, and if you break your leg nearer to a shore pine than a doctor, peel off a section of the bark for an emergency cast. In Asia pines are revered as a symbol of longevity. The fifth-century Chinese hermit poet T'ao Ch'ien wrote of returning from a long journey happy to be back in his garden, "walking around my lonely pine tree, stroking it."

Between the parking lot and the beach near Spanish Bank Creek • E side of Cambie St from 49th Ave S, next to the golf course.

Tips To Identify Pines

- needles in bundles (fascicles) of 2, 3 or 5
- "white" or "soft" pines in bundles of 5, cones usually cylindrical
- "yellow" or "hard" pines in bundles of 2 or 3, cones usually egg-shaped, scale tips prickly
- woody cones with tough scales
- needles often dropped when 3 years old, earlier than other conifers and creating a sparser outline

3 ▶ *Pinus nigra*
black pine

LEAVES Evergreen needles in bundles of two; up to 15 cm long and slightly twisted; dark green.

FORM Up to 40 m tall, rounded pyramid with thick, horizontal branching; dark; more imposing than lovely.

BARK Brown to grey with vertical fissures.

FLOWERS/FRUIT Cones in ovals, 5–10 cm long; scales have prickly tips.

European tree planted widely in North America, in part due to a deserved reputation for toughness. They don't mind air pollution and grow as well next to a busy street as they do in a quiet glen. The most commonly planted pine in Vancouver. Also known as Austrian pine.

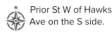

Prior St W of Hawks Ave on the S side.

Pinus ponderosa
ponderosa pine

LEAVES Evergreen needles in bundles of three (sometimes two); 12–28 cm long with sharp tips and edges; on tufts near the end of branches; dull green.

FORM Tall, straight trunk to 50 m with a diameter of 2 m; broad, open crown.

BARK Dark brown to black and scaly on young trees, maturing to thick, reddish-orange, flaky plates.

FLOWERS/FRUIT Seed cones 7–14 cm ovals with no stalk; scales get thicker towards the tips and have a sharp prickle.

An iconic evergreen of interior BC where deep roots help it thrive in dry areas. Thick bark also helps it outlast forest fires. The "ponderous" name refers to the massive size. First Nations uses include eating the seeds and inner bark, crafting dugout canoes from the trunk and making medicine from the sap to treat wounds.

At UBC on West Mall N of University Blvd • in Stanley Park on Park Lane S of the parking lot across the road from the tennis courts.

3 *Pinus sylvestris*
Scots pine

LEAVES Evergreen needles in pairs, 2.5–7 cm long and twisted; blue-green.

FORM Medium-sized tree to 15 m, often on a crooked trunk with an irregular, rounded crown that may flatten in maturity.

BARK Scaly, orange-red that darkens with age.

FLOWERS/FRUIT Seed cones 2.5–7 cm ovals, often in clusters of two or three pointing back against the stem.

The most widely distributed pine in the world, and one of the first pines to be planted in North America (in the 1750s). In some places it has naturalized (escaped cultivation). Varieties range from short and shrubby to tall and gnarled-branch beauties.

 In the median of King Edward Ave between Macdonald St and Vine St • in Arbutus Park at Arbutus St and SW Marine Dr.

Ginkgo biloba—ginkgo

4

Ginkgo biloba
ginkgo

LEAVES Thin and fan-shaped, often with a slight split into two sections (bi-loba like the name); 5–10 cm wide; bright green turning to yellow or gold in fall.

FORM Open and airy pyramid with sparse horizontal branching for several decades, later filling out majestically from a huge trunk; to 30 m tall.

BARK Pale grey, rough.

FLOWERS/FRUIT Apricot-like fruit on female trees contains a butyric acid that has been likened to rancid butter, but that's the polite description of the smell; nevertheless the nut inside is delicious roasted and has been used medicinally in Asia for centuries.

Said to be the oldest tree species on earth. The ginkgo we see today is virtually unchanged from fossils of 200 million years ago, meaning it lived through and beyond the age of the dinosaurs. There is nothing on Earth like it, and the genus has no surviving relatives, so it truly is unique. Researchers debate if it went extinct in the wild and was kept alive only by temple gardeners in Asia. Young ginkgo look gangly for years but turn into massive stunning beauties…eventually. In ancient China ginkgo was sometimes called the "grandfather-grandchild tree" for the three generations needed to see it mature. The venerable ginkgo is a symbol of longevity and resilience. Some temples in Japan nurture sacred specimens dating back more than 1,500 years, and six ginkgo trees famously survived the atomic bombing of Hiroshima even as buildings around them were obliterated.

 E Pender from Carrall St to Main St in Chinatown where savvy collectors gather the dropped fruit from the female trees in autumn.

Acer rubrum—red maple

Fraxinus excelsior 'Jaspidea'—Jaspidea ash

5 *Acer campestre*
hedge maple

LEAVES Leaves with 3–5 softly rounded lobes, 5–10 cm long and wide on a long stalk; dark green above, lighter beneath, nice yellows or sometimes reds and oranges in fall.

FORM Smallish, rounded crown to 10 m tall.

BARK Grey, ridged and furrowed.

FLOWERS/FRUIT Small green flowers in clusters, fruit a nearly horizontal samara as in Norway maple.

Also called field maple (a strange name for any forest tree), as well as common maple, English maple and small-leaved maple. Planted widely in decades past as a street tree that won't grow big and unruly.

Many along streets citywide including downtown such as the dozen or so on E Cordova St between Carrall St and Main St.

5 *Acer circinatum*
vine maple

LEAVES Roundish with 7–9 lobes and the same number of veins; green above, paler beneath, turning to red or yellow in fall.

FORM Multi-stemmed vine-like shrubs or small trees.

BARK Thin, green, maturing to darker reddish-brown but remaining smooth.

FLOWERS/FRUIT Drooping purple and white flower clusters appear in spring on branch tips; clusters of winged samaras 2–4 cm long spread at a wide angle.

A beautiful native tree not often planted outside our region, although it deserves to be. Vine maples are rainforest understory trees that send out long swooping branches in search of illumination. Where they find it, the leaves can take on an inner glow like nature's own spotlighting. As a maple they typically reproduce from winged seed samaras that helicopter away in a breeze, but they can also "walk" through the forest to a new site when a nodding branch tip touches the ground and takes root.

Sometimes used as a street tree such as the large one at 716 E 15th Ave • see them in their natural habitat of forested areas throughout the city such as Stanley Park and Pacific Spirit Park.

5 ▶ *Acer griseum*
paperback maple

LEAVES Three leaflets per leaf; green above, whitish-grey beneath, turning to bronze or red in fall.

FORM Small, oval to rounded with slender upright branching; 6–10 m tall.

BARK Beautiful cinnamon-orange, peeling away but remaining on the trunk in thin sheets to expose tan-brown bark beneath.

FLOWERS/FRUIT Winged samaras 3 cm long with large seeds.

As with any maple it puts out pleasing foliage, but the bonus comes in the unusual bark. Even, or perhaps especially, in the dormant season it's worth seeing for the vivid, peeling, papery skin. Native to central China as an understory tree in mountain forests, it has lately become popular here among garden designers looking for smallish trees with year-round interest.

Douglas Park at Willow St and W 22nd Ave N of the community centre.

5 ▸ *Acer macrophyllum*
bigleaf maple

LEAVES Familiar maple leaf shape with five deep lobes but growing bigger than a platter; dark green on top, paler underneath; stalks emit milky sap when cut.

FORM Largest maple species in Canada, up to 35 m tall with equal spread; narrow crown when grown in the forest with other trees, more spreading in the open.

BARK Smooth and brown when young, maturing to paler grey with scaly ridges.

FLOWERS/FRUIT Flowers in spring in drooping green-yellow clusters; winged samaras are 3–6 cm long and hairy.

A striking native maple that's a marvel for size, it earns its common name for leaves that resemble the Canadian flag's sugar maple on steroids. Not as colourful as the national icon maple in fall, but some years may still produce attractive yellows and oranges. Called the "paddle tree" in some Coast Salish languages for its hard wood deemed prime for making paddles and tools.

 See the Stanley Park Tree Tour map to find the largest maple tree in Canada • also 5672 Maple St along with others on the same appropriately named street.

5 *Acer saccharinum*
silver maple

LEAVES Maple leaf shape with five deep lobes and coarse teeth; dark green above, silvery-white (as in the common name) beneath.

FORM A long trunk with ascending branches, medium to large size up to 35 m.

BARK Smooth grey turning to shaggy with long, narrow flakes.

FLOWERS/FRUIT Earliest maple to bloom here with small yellow-green flowers in late winter.

An implant from eastern North America but not widely planted here. Naysayers find it grows too big too fast with brittle branches prone to failure and an aggressive root system that can clog sewers. May be tapped to make maple syrup, although not as prodigious as its more famous relative the sugar maple.

A beautiful five-trunked tree graces the SE corner of Oak St and W 28th Ave.

5 *Acer saccharum*
sugar maple

LEAVES Medium-sized, 8–15 cm wide with five lobes; yellowish-green above, paler beneath with brilliant oranges and reds in fall.

FORM Medium to large size up to 35 m tall, with a straight trunk, sturdy short branches and narrow rounded crown.

BARK Smooth grey becoming darker with vertical ridges.

FLOWERS/FRUIT Flowers small and greenish-yellow in drooping clusters before leaves emerge in spring.

The tree honoured everywhere a Canadian flag flies, but not planted much here as it doesn't seem to appreciate our wet Pacific climate. Thanks to one of the rare occasions where settlers learned from First Nations people how to live with trees, we now have maple syrup. Regrettably, we don't get the stretch of freezing nights and warm days typically needed to get enough sap flowing in spring, although some tap anyway. Syrup-heads report bigleaf maple and some birch trees can be used as substitutes, although you must collect more sap to get a similar amount of product. Distinguish sugar maple from the similar Norway maple (*Acer platanoides*), which we have along streets in droves, by seeing how Norway maple leaf stalks can grow longer than the blades and exude a milky sap when cut.

2700 E Broadway and Slocan St in front of the church together with a European hornbeam (*Carpinus betulus*).

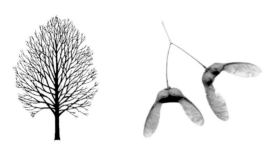

5 *Acer palmatum*
Japanese maple

LEAVES Small, star-shaped with 5–9 deep lobes and toothed edges.

FORM Multi-stemmed, upright or vase-shaped with layered branching, elegant; 6–10 m tall.

BARK Smooth, grey.

FLOWERS/FRUIT Small wide-spreading winged seeds to 2 cm across.

Native to Japan, China and Korea but now loved around the world for their architectural appeal and delicate foliage, it's one of the most widely cultivated trees in horticulture. You can join a fan club to try to sort out the different types, of which there are hundreds.

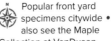

Popular front yard specimens citywide • also see the Maple Collection at VanDusen Botanical Garden.

LEAVES Broad and thin with 5–7 lobes that are more pointy than the sugar maple they resemble; dark green above and similar below, may show brilliant yellow or orange in fall.

FORM Round-topped crown to 20 m; ascending branches.

BARK Smooth to tightly ridged when mature.

FLOWERS/FRUIT Yellow-green flowers in spring before the leaves; nearly horizontal seed wings (samaras) are a quick way to tell Norway maple apart from related species.

The leaf looks enough like a sugar maple that when the new Canadian currency was issued in 2013, critics said the maples in the translucent windows were from this foreign interloper. (A government bureaucrat denied the claim, calling the leaf on the bills a "stylized" version of various indigenous species.) As a European introduction this tree gets flak for having been planted too widely. Back east it has caused problems by naturalizing to dominate native forests, but that hasn't been an issue here, and Norway maple continues to be selected as a tough but well-behaved civic tree. Among the many cultivars are Crimson King with dark purple leaves, Globosa which resembles a lollipop and strikes some as ridiculous, and Columnare, which urban planners like because its upright branches don't block traffic.

 1488 W 46th Ave just east of Granville St has three in a row • also on streets in nearly every neighbourhood citywide.

5 *Acer pseudoplatanus*
sycamore maple

LEAVES Thick, five-lobed, with toothed margins; to 14 cm wide and long; dark green above, paler underneath.

FORM Short trunk leading to a broad, arching crown; to 20 m tall.

BARK Greenish-brown and smooth when young, maturing to flaky brown with varying shades from grey to tan.

FLOWERS/FRUIT Drooping greenish-yellow clusters appear after the leaves; winged samaras 3–5 cm wide, held at a narrow angle.

European import widely planted in North America as a shade tree. Similar to Norway maple, but with rougher toothed leaves, samaras at tighter angles and leaf stalks that don't emit milky sap. *Pseudoplatanus* means "false plane tree" after another type it resembles.

341 W 20th Ave with others on the same street between Columbia St and Yukon St.

5 *Acer rubrum*
red maple

LEAVES Three-lobed, 5–15 cm long and wide; bright green above, paler underneath.

FORM Medium-sized to 25 m tall with ascending limbs making a dense crown.

BARK Smooth light grey maturing to darker grey with scaly ridges.

FLOWERS/FRUIT Small red flowers in clusters; seed wing samaras to 3 cm long with narrow angles.

Native to eastern North America and planted widely here as a street or park tree for spectacular visuals from early spring to late fall. Earns the name *rubrum* for beautiful red tones through the seasons, moving from buds to twigs to flowers to seeds to leaves. Plenty of variations in both nature and in cultivation where the named versions describe appealing attributes such as Red Sunset and Embers.

A well-known double row of Red Sunset cultivars frame the sidewalk beside the law courts on the E side of Hornby St between Robson St and Georgia St—architect Arthur Erickson was reported to have wanted London plane trees but was overruled by the city • also a row of 7 Red Sunset cultivars on the Memorial Park South side of Ross St from 45th Ave N.

5 *Aesculus hippocastanum*
horsechestnut

LEAVES Compound with 5–9 (usually 7) leaflets on a long stalk; each leaflet 10–25 cm long with middles larger than the ends; green turning yellow-brown in fall.

FORM Robust medium to large tree with thick trunk, spreading crown, to 25 m tall.

BARK Grey, smooth when young, maturing with fissures and many scales.

FLOWERS/FRUIT White, bell-shaped, 2–3 cm long, in cone-shaped clusters that look like someone stocked the tree with upright candles in spring; fruit in 5- to 6-cm-wide spiked green globes which break open to reveal shiny and beautiful, alas inedible, nuts.

Planted all over in parks and as street trees. The brown nuts in fall sometimes fool people into thinking they've found a lode of delicious chestnuts. Unfortunately they contain saponin, a toxin that would need to be leached out in a laborious process of cooking and rinsing which would also presumably remove the nutrients and taste— anyway no one bothers. They can, however, be used for the fine schoolyard game of conkers, which offers all the fun of a video combat game without the video or the violence.

E 10th Ave between Carolina St and St. George St along the bike route • also on streets and in parks all over the city.

5 *Catalpa speciosa*
northern catalpa

LEAVES Heart-shaped with pointed tip; very large, to 30 cm long; pale green turning yellow in fall.

FORM Medium to large tree with muscular branching, to 30 m tall.

BARK Dark brown, irregular fissures.

FLOWERS/FRUIT Large clusters of flowers with a tubular base, white with yellow and purple spots; fruit hangs in 35-cm-long cylindrical pods, green turning to brown.

Larger than its cousin southern catalpa (*Catalpa bignoni-oides*, which is also called common catalpa or cigar tree) and growing more upright, with bigger leaves showing longer tips. Northern catalpa also flower earlier and develop thicker bean pods. Planted beyond their native southeastern US for the showy flowers and attractive form even in winter. The huge leaves add a remarkable appeal. You can gauge the size potential of a golden catalpa (*Catalpa bignonioides* 'Aurea') by checking the one planted on August 30, 1975, to commemorate the opening of VanDusen Botanical Garden. Northern catalpa is also known as western catalpa or common catalpa.

10th Ave from Stevens St to Blenheim St • Thornton Park at Main St and Terminal Ave.

Catalpa bignoniodes 'Aurea'

5 *Cercidiphyllum japonicum*
katsura

LEAVES Heart-shaped or nearly round, to 7 cm long with three main veins; colourful reddish-purple in spring, then green in summer, then golden yellow in fall.

FORM Pyramid maturing into a broad-spreading crown to 20 m or taller; often multi-trunked.

BARK Grey to brown, furrowed on mature trees.

FLOWERS/FRUIT Small red tufts in early spring before leaves; male and female flowers on separate trees.

Native to China and Japan. A gorgeous tree for the yard, park or wherever it can fit. As a bonus in fall, the golden leaves scent the air with a fragrance as if a toffee factory opened next door. The Pendula cultivar can enhance the considerable beauty of this tree with cascading streams of leaves. It's easy to fall in love with katsura trees.

Mature trees on Brightwood Pl at Vivian Dr across from Fraserview Golf Course entrance • E 3rd Ave from Skeena St to Cassiar St.

Cercidiphyllum japonicum 'Pendula'

Fall colour

5 *Cornus kousa*
kousa dogwood

LEAVES Pointed, elliptical, 4–6 pairs of parallel veins; 5–10 cm long; dark green above, hairy below.

FORM Tall shrub or small tree to 7 m with rounded top.

BARK Thin, orange-grey, scaly when mature.

FLOWERS/FRUIT Tiny flowers surrounded by four 5-cm-long pointed white bracts which give the appearance of a larger flower; fruit held upright on stalks in 2–3 cm soft, reddish, knobby balls.

Dogwoods are pretty trees that have inspired growers to develop dozens of cultivars. Kousa has recently gained popularity because it resists the anthracnose fungus that has wrecked havoc on our native Pacific dogwood (*Cornus nuttallii*). It flowers in late spring after other dogwoods are finished, providing a delightful display that can cover the branches. Also produces a kind of fruit in upright, spikey, soft red globes that are edible and sweet, though the mealy texture may convince you to leave the rest on the tree. Also known as Japanese dogwood or Korean dogwood.

Traffic circle at Charles St and Cotton Dr.

LEAVES Pointed, elliptical, with typical dogwood parallel veins, 5–10 cm long; dark green, hairy above and below.

FORM Small trees or tall shrubs, usually multi-stemmed.

BARK Brown-grey to dark brown, shedding on maturity.

FLOWERS/FRUIT Yellow flowers and bracts in spring before the leaves; fruit small, cherry-shaped, bright red, edible although they may be astringent until very ripe.

Native to Europe and Asia. Planted here less for the fruit, which may or may not be worth collecting depending on the tree and that year's growing conditions, and more for the cheery yellow flower clusters in early spring when little else is blooming.

At the SE corner of the City Hall grounds at 10th Ave and Cambie St.

5 *Cornus nuttallii*
Pacific dogwood

LEAVES Oval with pointed tip and veins roughly parallel to the leaf edge; to 10 cm long.

FORM Small tree or large shrub, multi-stemmed, irregular crown; to 20 m.

BARK Smooth, dark grey, becoming ridged with age.

FLOWERS/FRUIT Attractive white "flowers" are actually modified leaves that surround the tiny greenish-white real flowers; fruit in berry-like 1 cm balls of bright red, favoured by grosbeaks, thrushes and other birds.

Not the provincial tree (that's western redcedar) but it does contribute the official flower. It appears on the provincial shield above the motto *Splendor Sine Occasu* which is also a mouthful in the English translation "Splendour Without Diminishment." The bark has traditionally been used in medicine to purify blood. The wood is good for bows, arrows, tool handles and hooks, but is not harvested commercially—there's a law against killing this tree in BC. Sadly a fungal disease, anthracnose, has taken out many of the city's trees. A Vancouver original cultivar, Eddie's White Wonder, is a

cross between our native dogwood and the eastern flowering dogwood. Nurseryman Henry Eddie bred the new tree in 1945 which became a sensation and is now planted worldwide.

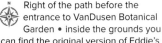

Right of the path before the entrance to VanDusen Botanical Garden • inside the grounds you can find the original version of Eddie's White Wonder.

LEAVES Compound leaf with 9–15 leaflets, each 5–8 cm long and sharply toothed; dark green turning bright yellow in fall.

FORM Broadly spreading shade tree growing up to 40 m tall and wide, although usually smaller here.

BARK Smooth, pale grey when young, becoming thick and vertically fissured.

FLOWERS/FRUIT Seeds contained in wide, flat, slender samaras with only one wing, a good way to identify an ash tree.

European native planted often in North American parks. There are many cultivars of varying shapes, usually recognizable by their compound, oppositely-arranged leaves. Ash trees have played a long and sometimes sacred role in history, including serving as Yggdrasil, the tree of life in Old Norse mythology. The wood is excellent for tool handles as it absorbs shock; it also burns well, even when green. For a winter ID tip look for stout greenish-grey shoots with very black buds.

 Cambie St and 29th Ave in the NW corner of Queen Elizabeth Park.

5 *Fraxinus latifolia*
Oregon ash

LEAVES Compound in 5–9 but usually 7 leaflets, each 4–10 cm long and broader than on many ash trees (*latus* is Latin for broad, *folia* for leaf).

FORM Medium to 50 m tall with narrow crowns when forest grown, more spreading in the open.

BARK Reddish-brown to grey with deep fissures and scaly plates.

FLOWERS/FRUIT Greenish-white flowers in hanging clusters; fruits are elliptical samaras with single keys, 2.5–5 cm long.

Botanists aren't certain whether the native range of Oregon ash goes as far north as BC or if the tree has naturalized here. It isn't often planted in gardens, perhaps because it's considered native and thus too common for acclaim, but it should be since it grows well here, looks great and puts on a nice fall colour.

Trout Lake (John Hendry Park) between the west side parking lot and the water.

5 *Paulownia tomentosa*
princess tree

LEAVES Oppositely-arranged, very large (leaves on new shoots from the trunk can be 50 cm or wider), heart-shaped to mildly lobed; downy green.

FORM Medium to large tree, pyramidal to irregularly rounded; grows to 20 m in just a few years.

BARK Smooth and grey with broad ridges (not dark and fissured like the similar-leaved catalpa).

FLOWERS/FRUIT Light purple, fragrant and tasty flowers in 20-cm-long clusters on branch tips in spring before the leaves emerge; seed pods in small, pale green capsules that ripen to brown and stay on the tree in winter.

An Asian tree now naturalized in parts of eastern North America. Some claim this started when dried seed pods were used as throw-away protective packing around shipments of porcelain from China. Certainly they don't need coaxing to grow. A forgotten seedling in the corner of a garden can tower above the house in two or three years. Invasiveness doesn't seem a problem here where *Paulownia* are admired for their gorgeous spring flowers. A Japanese tradition held that the birth of a baby girl should be honoured by the planting of this tree known as *kiri*; by the time she was old enough to marry it would provide enough wood to build a marriage chest for her dowry. Named after Anna Pavlovna, daughter of the czar, it is also called royal paulownia, empress tree and foxglove tree.

Thornton Park in front of the Central Railway Station on Main St and Terminal Ave where you can compare to the similar looking but unrelated catalpa: *Paulownia* have leaves that are larger and arranged in opposite pairs (not whorls), smoother bark, purple (not white) flowers earlier in spring and rounded, woody fruit capsules rather than long bean pods.

Platanus x acerifolia—London plane trees

6 *Ficus carica*
fig

LEAVES 3–5 deep lobes, to 20 cm long and wide; dark green.

FORM Multi-stemmed spreading tree, may reach 25 m tall and wide.

BARK Smooth, thin, grey.

FLOWERS/FRUIT The familiar and delicious "fruit" is actually a syconium made of tiny flowers growing inside the skin.

Figs have fascinated and fed gardeners for thousands of years. The more you read, the weirder they get. Native to eastern Mediterranean hot spots like Sicily and Turkey, they survive our typically mild winters and, depending on the amount of summer light and heat, may produce abundant crops. Unlike figs in some areas, which must be pollinated by a particular type of wasp, the edible types grown here need no pollination. Fig trees overall are amazing. A bo tree (*Ficus religiosa*) in Anuradhapura, Sri Lanka, grown from a cutting of the tree under which Buddha found enlightenment, was planted in 288 BC and is still revered there today.

E 64th Ave just E of Main St • or just look in yards and back lanes citywide, especially where Mediterranean immigrants have settled.

6 *Liriodendron tulipifera*
tuliptree

LEAVES Four (sometimes six) lobes with a flattened top; 7–14 cm long, held on a long stalk; bright green above, paler underneath, yellow in fall.

FORM Tall, straight single trunk with arching side branches and ascending twigs; to 30 m or more.

BARK Smooth and faintly striped grey when young, maturing to darker grey with deep furrows.

FLOWERS/FRUIT Large, tulip-shaped, yellow-green flowers emerge after the leaves; cone-like fruits produce samaras (winged seeds).

Stately member of the magnolia family, native to eastern North America but planted far beyond. Can get big—it's the tallest deciduous tree on the continent. Named after its tulip-looking flowers, but you may not see them up in the lofty crown so the easiest ID feature is the distinct four-pointed leaves. A favourite of aphids, which don't seriously harm the tree but do annoy those who park their cars underneath to find them covered

Douglas Park north of the community centre at Willow St and W 22nd Ave has one beside the playground that needs three adults with interlocked arms to reach around the trunk.

with sticky honeydew. Also known as tulip poplar, yellow poplar and whitewood. The smaller Chinese tuliptree, *Liriodendron chinensis*, has recently been planted in various places around the city; you can find examples at VanDusen Botanical Garden in the Sino-Himalayan area.

6 *Liquidambar styraciflua*
sweetgum

LEAVES Star-shaped with 5 or 7 lobes; green above, paler beneath, colours from yellow to purple in fall; resin smell when crushed.

FORM Narrow pyramid to 25 m tall.

BARK Smooth grey maturing to rough grey-brown with ridges.

FLOWERS/FRUIT Single female flower is a 1 cm ball on 5-cm-long stalk; male flowers in 7-cm-long terminal spikes; fruit 2.5-cm stalked woody balls that hang on through winter.

Eastern North American tree planted here for its pleasing form and fantastic fall colours. The leaves may look like some kind of maple but you won't make that ID error because you'll recall maples have leaves in opposite pairs. The Worplesdon cultivar lacks the gumball fruit which some find messy; it has recently gained popularity as a street tree.

 Along Nicola St and Cardero St at Coal Harbour are a number of Worplesdon sweetgums.

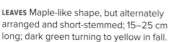

LEAVES Maple-like shape, but alternately arranged and short-stemmed; 15–25 cm long; dark green turning to yellow in fall.

FORM Massive trunk and thick, wide-spreading branches; pyramidal when young maturing to rounded crown; to 35 m tall and wide.

BARK Peeling camouflage-like patchwork of cream, tan and grey with lighter bark beneath.

FLOWERS/FRUIT Male flowers small, yellow, clustered along twigs; seed flowers crimson, on long stalks; fruit hangs in 2- to 3.5-cm balls on 8–16-cm-long stalks through winter, typically in pairs (fruit from one of its parents, American sycamore (*Platanus occidentalis*), usually hangs singly).

This cross between American sycamore and Oriental planetree became popular in London in the 1800s as a hardy urban tree that could handle polluted air and poor soil. Half that city's trees are said to be London planes and it has been planted from New York to Paris as a prominent feature in parks and along streets. The cultivar known as Pyramidalis is sometimes used to fit into tight urban spaces a London plane would overwhelm. Not all experts agree with that cultivar's classification—with hybrids going back to the 1600s in England it can be hard to pin the many varieties down. Also known as sycamore, hybrid plane or hybrid sycamore.

Heatley Ave from Prior St to Atlantic Ave • and all over.

6 ▶ *Quercus garryana*
Garry oak

LEAVES Thick and stiff, 5–7 lobes, 7–10 cm long; dark green above, duller green underneath.

FORM Small to medium tree to 20 m tall with broad, rounded crown made of many twisted, gnarled branches.

BARK Grey and furrowed into sections with maturity.

FLOWERS/FRUIT Acorns 2–3 cm long held in shallow cups; edible but you may need to soak to remove tannins.

The only native oak in BC, although seen more often in drier areas of the Gulf Islands than in wet Vancouver. Important enough to have a native ecosystem named after it, alas one that's threatened as urban development eats up ever more Garry oak meadows. Named by plant hunter David Douglas after Hudson's Bay Company secretary Nicholas Garry who helped him in his quests. Also known as Oregon white oak.

Two in front of 1529 W 71st Ave W of Granville St and one on the W side of Dunbar St between 19th Ave and 20th Ave show why native plant proponents love this tree.

6 *Quercus palustris*
pin oak

LEAVES Thin, 5–7 lobes with some bristle tips on larger lobes and deep, wide U-shaped notches; 7–15 cm long; shiny dark green above and paler beneath, red in fall; dead brown leaves may hang on the tree in winter.

FORM Medium-sized trees to 20 m tall; trunk often straight to the top, keeping a pyramidal shape, unlike most oaks which get spreading crowns; numerous slender, drooping branches.

BARK Smooth brown, shallow fissures when mature.

FLOWERS/FRUIT Drooping catkins among the first from oaks to flower in spring; acorns squat, about 1 cm long, held in shallow cups with tight scales.

Native to eastern North America where it's found along waterways (the name *palustris* means "from swamps"). Planted often as a street tree back east because it tolerates both flooding and drought, and the relatively narrow form doesn't affect traffic.

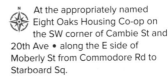

At the appropriately named Eight Oaks Housing Co-op on the SW corner of Cambie St and 20th Ave • along the E side of Moberly St from Commodore Rd to Starboard Sq.

LEAVES Thin, 3–7 rounded lobes with two smaller lobes at the base; 5–12 cm long; dark, dull green above, paler beneath.

FORM Short trunk, thick branches in a rounded to irregular crown, can reach 35 m tall and wide.

BARK Dark grey with deep fissures.

FLOWERS/FRUIT Acorns to 4 cm long in clusters of 2–5, on long stalks.

Oak trees were legendary in European history from tree-worshipping druids to Robin Hood's band in Sherwood Forest. In Norse mythology the oak was sacred to Thor, the god of thunder, who was so revered he got his own Thor's Day, which we call Thursday. Used to build much of the British Navy and brought to North America with settlers. The most commonly planted oak here, it has naturalized in forests such as Stanley Park and Pacific Spirit Regional Park. Also known as British oak or common oak.

Ten of the more slender pyramidal English oak (*Quercus robur* 'Fastigiata') were planted in 1999 on Columbia St between 46th Ave and 47th Ave; "tragedy" and "comedy" oaks inspired by writing from the Bard are planted in the Shakespeare Garden in Stanley Park.

6 *Quercus rubra*
red oak

LEAVES Leaves with 7–9 deep, pointed lobes and bristle tips on larger lobes; yellow-green with paler undersides in summer, then red or orange in fall.

FORM Straight, thick trunk with upright branching and symmetrical rounded crown when older, to 30 m tall.

BARK Smooth, dark grey with long fissures.

FLOWERS/FRUIT Flowers in drooping yellow-green catkins; acorns to 2.5 cm long and nearly as wide, held in a shallow cup.

Oaks come in many types with more than 600 species ranging from shrubs to majestic trees. Red oak is native to the forests of eastern North America, planted here as an attractive medium to large tree with fine fall colour. The original Vancouver PoetTree was designated as the spectacular red oak at the corner of Beach Ave and Bidwell St in Alexandra Park across from the inukshuk at English Bay. Canada's Poet Laureate Fred Wah inaugurated the concept of providing a place for people to express themselves poetically in 2013 by reading his poem *Tree*.

Turner St from Kootenay St to Skeena St is lined with two dozen tall red oak trees. • The Main Mall at UBC has a fine row of red oaks, some planted in the 1930s.

Original Vancouver PoetTree, a *Quercus rubra*—red oak in Alexandra Park.

Tree

Go to the forest, Tree
please wait for me there

Need to be seen
cold (not) but if the roots

Hold a world that floats beneath
what holds us

Our floor covered dry, deciduous
shrapnel scattered at our feet

Goodbye, but also, silently hello
we will say.

— *Fred Wah*

Planting of new Vancouver PoetTree, a *Catalpa bignoniodes* 'Aurea'—golden catalpa, at Trout Lake.

7

Alnus rubra
red alder

LEAVES Thick, oval with pointed tip, roughly toothed with 8–15 veins on either side of central vein; 7–13 cm long; green above, greyish underneath.

FORM Medium-sized to 25 m tall; straight and narrow crown; branches to the ground when grown in the open.

BARK Smooth, light grey, separates into plates with age; exposed underbark turns orangish-red, hence the name.

FLOWERS/FRUIT Reddish drooping male catkins appear in late winter, turning to yellow in spring; female catkins in clusters of 3–5 turn woody and stay on tree through winter.

Red alder doesn't get a lot of love in the city where it's probably the most numerous native deciduous tree. No one plants it because it's likely to show up anyway in a newly exposed plot, being an early adopter. Even in poor soil, nodules on the roots help fix nitrogen, which prepares the site for eventual takeover by longer lasting conifers such as hemlock and Douglas-fir. The colourful underbark can be used to dye baskets, clothes and hair. The inner cambium layer is a traditional food source. Try using alder leaves as insoles on a hike to comfort aching feet. Alder wood is good for underwater construction because it resists rot when wet—Venice is said to be built on pilings of alder. Feel the rough underside of the leaf edge to confirm ID; alder leaves have slightly rolled edges.

The second largest red alder tree in Canada is just to the N of the concession stand at Third Beach in Stanley Park— a near-record tree almost no one notices.

7 *Betula papyrifera*
paper birch

LEAVES Triangular with a pointed tip and irregular teeth; 5–10 cm long; dull green above, paler beneath.

FORM Small to medium, to 30 m tall with a slender, sometimes curved trunk, often multi-stemmed; numerous ascending branches create irregularly oval crown.

BARK Thin, smooth, reddish-brown when young, maturing into white, papery sheets with irregular rough sections; peels off easily, but don't strip a tree just to find out what it is, or was.

FLOWERS/FRUIT Male catkins hang in clusters of 1–3 through winter.

Native to our region and throughout forests across Canada. The bark is used to make everything from containers to canoes and the wood is carved into spoons, masks and more. If you're allergic it may not please you to learn that a single catkin can hold 5.5 million grains of pollen. Birch bark contains a powdery substance called betulin, which not only has waterproof properties but has been shown to help wounds heal and reduce inflammation. In the Old World birches were believed to have magical properties, thus baby cradles were often made out of birch as a form of protection. Also known as white birch or canoe birch.

 156 E 19th Ave across from the park.

LEAVES Triangular with narrow tip and double-toothed edges; 3–7 cm long on a relatively long stalk.

FORM Small to medium trees to 15 m tall, often multi-stemmed; upright branches holding thin, droopy twigs in a broad crown.

BARK White and peeling in strands; rough, diamond-shaped black furrows near trunk base.

FLOWERS/FRUIT Seed catkins 2–4 cm long, blunt-tipped, hanging on stalks.

Native to Europe and Asia, popular as a street tree here in years past but less favoured now since a beetle known as the bronze birch borer has damaged or killed a number of ornamental birches. The pest problems are worse for stressed trees, a good description for many urban birches that would rather live in a cool, moist forest. Also known as weeping birch or silver birch.

Several attractive ones on the S side of Union St between Jackson Ave and Dunleavy Ave.

7 *Carpinus betulus*
European hornbeam

LEAVES Oblong to a narrow tip and sharply toothed with 10–13 pairs of prominent veins; 5–10 cm long.

FORM Upright, stately pyramid with slender branching, grows less regular with age; to 20 m tall.

BARK Smooth, pale grey.

FLOWERS/FRUIT Chains of hanging nutlets with three wings, the middle being the longest, turning from green to brown.

Finely-shaped European tree introduced to cities elsewhere because it's both pretty and hardy. The cultivar Fastigiata which grows upright into a tight oval, avoiding unplanned pruning from trucks and buses, is seen fairly often here.

E side of Renfrew St from Parker St to William St; the N side of Prior St from Campbell Ave to Gore Ave has a long row of Fastigiata.

LEAVES Oblong with a rounded base narrowing to a sharp tip and bold forward-pointing teeth; 16–28 cm long.

FORM Thick trunk holding upright branches in rounded crown, to 35 m tall.

BARK Smooth, dark brown, developing deep, spiral furrows and flat-topped ridges.

FLOWERS/FRUIT Male flowers hang in white catkins 10–20 cm long; spiny burred shells hold edible nuts, for trees that get pollinated.

Native of southern Europe, introduced centuries ago to North America by the Spanish and now naturalized in eastern forests. The similar American chestnut (*Castanea dentata*) was once a native mainstay of eastern forests but both have been obliterated by chestnut blight, a fungal infection that has wiped out chestnut trees in most of the continent, although not here. Also known as Spanish chestnut or European chestnut.

 Two beauties in Kitsilano Beach Park, just N of the parking lot at McNichol Ave and Arbutus St.

7 *Corylus avellana*
common hazelnut

LEAVES Rounded oval with pointed tip and coarsely double-toothed, 6–12 cm long and across; green above, paler beneath, yellow in fall.

FORM Multi-stemmed shrub or spreading tree to 10 m tall.

BARK Smooth, greyish brown.

FLOWERS/FRUIT Long, hanging, yellow male catkins seen in winter and early spring; involucres (husks) don't extend beyond the nut (beaked hazelnuts have elongated involucres).

The hazelnut, also called a filbert, is the only nut grown commercially in our region. Orchards in the Fraser Valley used to export more than 300 tonnes a year, but production dwindled when the eastern filbert blight began killing trees. New blight-resistant varieties are now being grown. The ornamental Contorta cultivar known as corkscrew hazel or Henry Lauder's walking stick is derived from a single misshapen specimen found growing in a hedgerow in England in 1863.

 Four on the E side of Highbury St just N of W 1st Ave.

LEAVES Rounded oval with pointed tip and coarse double teeth, heavily veined, 5–10 cm long; green above, paler beneath, yellow in fall.

FORM Multi-stemmed shrub or spreading tree to 10 m tall.

BARK Smooth greyish-brown when young, later developing a crisscross pattern.

FLOWERS/FRUIT Male flowers hang in long catkins before leaves emerge in spring; female catkins small with red tufts; nuts enclosed in husks with a tubular extension that reminds some of a beak.

The californica variety of beaked hazelnut is a native from here down to California, known as a valuable food source for First Nations. Indigenous hazelnut farmers would selectively burn production areas to encourage new growth. Difficult to distinguish from the introduced common hazelnut trees when the nuts aren't available, but beaked hazelnuts often have catkins with one or two in a cluster on a very short peduncle (stalk); common hazelnuts usually have catkins with more than one in a cluster on 1-cm-long peduncles; they also bloom earlier than the native species.

Seen in forested areas and back lanes citywide as squirrels plant both the native beaked hazelnut and the introduced common hazelnut • the N side of Nelson St from Thurlow St to Bute St has the related Turkish hazel (*Corylus colurna*) which is grown for its looks rather than food.

7 *Crataegus x lavallei*
hybrid hawthorn

LEAVES Oblong with pointed tip and toothed edges; 4–10 cm long; glossy green above, paler beneath.

FORM Small to medium with dense branching in an oval crown; 4–8 m tall.

BARK Smooth, greenish-brown when young, breaking into scaly plates with age.

FLOWERS/FRUIT Small white flowers clustered along twigs; fruit 1 cm long, red, like a tiny apple but with just two seeds, hanging on trees in winter.

If you'd like a botanical puzzle to occupy the rest of your life try classifying all the various types of hawthorn. More than 1,100 species were listed a century ago; efforts to wrangle them into neater categories since then haven't really cleared up the confusion. In general terms, hawthorns are small to medium ornamental trees with pretty white to red flowers, dense crowns of twisting branches and tiny attractive fruit (haws) which appeal more to birds than people. They may also have thorns. *Crataegus* comes from the Greek word *kratos* meaning "strength" for its hard wood. The fruit, flowers and haws are also used in herbal medicines, particularly for the heart.

The S side of W 14th Ave from Cambie Blvd to Heather St has about a dozen planted in 2005.

7 *Davidia involucrata*
dove tree

LEAVES Heart-shaped, large-toothed, on a long stalk, to 18 cm long; dark green above, paler beneath.

FORM Short trunk and oval crown to 12–16 m tall.

BARK Smooth and grey, maturing to flaky orange.

FLOWERS/FRUIT Flowers in clusters with two white bracts 15 cm long drooping from branches; fruit (not edible) a 3-cm-wide hard greenish ball hanging on a 6-cm-long stalk.

Native to China. Not so widely planted in North America but a show-stopper in spring when flowers with creamy white bracts (leaves that look like flower petals) cover the tree like a flock of doves or, if you prefer, handkerchiefs. Garden writers often repeat the line that it's a drab tree when not in flower, but they're not looking closely enough.

The tall, spreading tree in a yard on SW Marine Dr at 49th Ave has delighted drivers for decades in spring • you can get closer to a nice one in Queen Elizabeth Park on the western stairway down from the main plaza.

7 *Fagus sylvatica*
European beech

LEAVES Oblong on an uneven base, with 5–9 pairs of veins, to 10 cm long; slender, pointed, brown buds are a winter ID clue.

FORM Stout trunk leading to multiple upright spreading branches in an attractive shape; to 30m tall.

BARK Pale grey, smooth even in maturity with a texture some dunces feel compelled to carve their initials into.

FLOWERS/FRUIT Male flowers inconspicuous green balls in spring; fruit in bristly husks covering edible (according to squirrels) nuts.

An elegantly large tree that earns its frequent role as a park standout—and it had better since the dark shade beneath makes anything else struggle to grow. Many cultivars in varying shapes and colours including copper beech (Atropunicea) which has purple leaves, fernleaf beech (Asplenifolia) with deeply cut lobes and Pendula or weeping beech. Distinguish European beech from the less-seen-here American beech (*Fagus grandifolia*) which has longer and narrower leaves with more veins and sharper teeth. Van-Dusen Botanical Garden has a mature Beech Collection to help you sort the various types out.

 14th Ave between Yukon St and Cambie St.

LEAVES Evergreen, waxy, spiny toothed; 3–7 cm long; glossy dark green.

FORM Often pruned to a shrub but can grow into a sizeable tree with a pyramidal crown 15 m or taller.

BARK Thin, smooth, grey.

FLOWERS/FRUIT Tiny, greenish to white flowers; bright red berries eaten and pooped out by birds, leading to its reputation as an invasive species.

Familiar sight from Christmas displays with bright red berries (on female trees) against dark green leaves. Hundreds of cultivars and hybrids offer many sizes and shapes and colours. Introduced from Europe but has escaped cultivation and is now naturalized (able to live and reproduce on its own) in forests in the Pacific Northwest.

The Holly Collection at VanDusen Botanical Garden is a good place to compare various types, including some that look nothing like the hollies you expect.

7 *Malus domestica*
apple

LEAVES Oval, pointed, finely toothed and rough-surfaced; 7–10 cm long; green above, paler and hairier beneath.

FORM Short, stout trunk and wide-spreading crown.

BARK Grey-brown, thick, scaly, rough.

FLOWERS/FRUIT Large, pink to white, appearing in spring before or with the emerging leaves on short, gnarled fruiting spurs.

Four species of *malus* are native to North America, but the apple we all love is a descendent from Eurasia, since domesticated into thousands of cultivars. Plant an apple tree and you may become enraptured by the species in all its aspects from history to culture to taste. Apples were important enough in the daily life of Western society to affect the language—we may still call a sweetheart "the apple of my eye" and a person who's unpleasant "crabby." Although Vancouver does not have ideal apple growing conditions—our dripping springs and summers can host a glut of pests and diseases—dedicated growers still manage to produce delicious fruit most years.

UBC Botanical Garden and VanDusen Botanical Garden both have apples on display • the Strathcona Community Garden W of Hawks Ave between Prior St and Malkin Ave has varieties that date back centuries.

7 *Malus fusca*
Pacific crabapple

LEAVES Egg-shaped with elongated tip and toothed, sometimes lobed; to 10 cm long.

FORM Multi-stemmed shrub or small tree to 12 m tall; sharp pointed spurs.

BARK Reddish brown with large, flat scales.

FLOWERS/FRUIT Small white-to-pink flowers in spring; fruit red or yellow with red patches, oblong, 1–2 cm long; edible if you like quite tart fruit.

Our only native apple tree, Pacific crabapples have long been an important First Nations food and medicine. Can be eaten fresh, preserved into jams and jellies or stored in the traditional way in water or oil. They get sweeter as they soften. Medicinal preparations from the bark may be used to treat upset bellies, rheumatism and dysentery. The durable wood is good for tool handles, bows, halibut hooks and more.

Hastings Park Sanctuary on East Hastings St E of Renfrew St has several Pacific crabapple trees among its native plantings • N of Bloedel Conservatory in Queen Elizabeth Park.

LEAVES Heart-shaped with narrow tip and sometimes irregularly lobed; boldly toothed and thick; to 20 cm long.

FORM Twisting trunk with vigorous upright branching; to 12 m tall and wide or wider.

BARK Reddish-brown, scaly.

FLOWERS/FRUIT Flowers insignificant; fruit (drupes) resemble blackberries as they emerge red and ripen to black; very tasty and juicy; to 4 cm long.

May be confused with white mulberry (*Morus alba*), which is grown as an ornamental (or historically to provide leaves for silk worms), and red mulberry (*Morus rubra*), which is native to central and eastern North America. Black mulberry comes from central Asia. The biggest difference is the taste: black mulberry can produce a fruit so delicious one berry may convince you to plant your own tree. If you really want fruit you almost have to—it's too tender and juicy to travel as a commercial product. Also known as Persian mulberry or common mulberry.

 A big and beautiful one in VanDusen Botanical Garden near the eastern end of the Rhodo Walk.

7 *Parrotia persica*
Persian ironwood

LEAVES Lopsided ovals with wavy or toothed edges, to 20 cm long; dark green turning yellow-orange-red in fall.

FORM Large spreading shrub or smallish tree with low branching and a rounded crown; to 12 m tall.

BARK Smooth grey when young, maturing to exfoliating grey revealing attractive patches of green, brown and white.

FLOWERS/FRUIT Small red flowers in late winter before leaves emerge; fruit 1 cm wide in a dry capsule holding two seeds.

Native to northern Iran but admired beyond its range as a pest- and disease-resistant tree that may get overlooked all summer, then turn heads in fall with spectacular colours. City planters admire its hardiness as well as looks. The flaky, multi-coloured bark is appealing in bare winter months.

Two planted in 2008 in front of 3035 W Broadway W of Carnavon St are the Vanessa cultivar which grows more upright than the standard • the Ruby Vase cultivar has recently been planted along commercial streets in various parts of the city, so it seems Persian ironwood has a solid future here.

LEAVES Round to heart-shaped with a pointed tip; stalks are longer than the leaf and flattened, making the foliage "quake" in a breeze; to 12 cm long; green above, paler beneath, yellow-gold in fall.

FORM Tall, graceful, slender with upright branching, to 24 m.

BARK Smooth, whitish-green, becomes darker and furrowed with age.

FLOWERS/FRUIT Catkins 4–6 cm long in spring before leaves emerge; fruit a 10-cm-long string of seed capsules.

The most common deciduous tree in Canada, and thriving throughout the northern forests, although not prevalent here on the coast. The dainty name and pretty looks may be deceptive. Quaking aspen are tough pioneer trees quick to establish themselves on exposed sites. Although fast-growing individual trees may last no longer than 50–60 years, they send out root suckers in all directions, so what looks like an aspen forest may genetically all be the same tree. The Pando grove in Utah is technically one tree with 40,000 stems covering 83 ha of land, and the roots are believed to be 80,000 years old. Also known as trembling aspen, Canadian aspen or American aspen.

 A nice compact grove in a large planter at the NE corner of Burrard St and Nelson St. • also a delightful grove in VanDusen Botanical Garden.

7 *Populous trichocarpa*
black cottonwood

LEAVES Broad oval with narrow tip; 7–12 cm long (larger leaves up to 30 cm long may appear on water sprouts and young trees); green above, paler beneath.

FORM Large with upright growth to 35 m or taller; thick trunks 1 m or more in diameter.

BARK Smooth, green when young, maturing to grey with fissures and flat-topped ridges.

FLOWERS/FRUIT Male flowers in catkins 4–6 cm long; female catkins 6–9 cm long release tiny seeds covered in white fluff that can fill the air like early summer snow.

A massive native tree, its wind-borne seeds and knack for growing in difficult conditions mean a newly cleared lot in Vancouver is likely to be head-high in cottonwoods a few years later. Cotton-wood is an important aboriginal tree with the inner bark used for both food and medicine. The shiny brown, pointed, sticky and fragrant buds in winter are an ID clue. If you boil them to release the sweet resin and mix it with olive oil and bees wax you'll have Balm of Gilead,

A row of tall ones at the S end of Strathcona Park next to Cottonwood Community Garden at Raymur Ave and Malkin Ave includes one that sometimes hosts a pair of nesting bald eagles.

good for soothing aches. Black cotton-wood is in the willow family (Salicaceae) which has been used since the ancient Sumerians in 2000 BC to reduce pain and fever (salicin from willow bark was turned into aspirin in 1897). Bees collect the resin from black cottonwood to make propolis, an anti-bacterial agent that protects the hives from invaders.

7 *Prunus avium*
sweet cherry

LEAVES Narrow oval with pointed tip and toothed, 10–12 cm long; dark green above, hairy beneath.

FORM Upright growth, often conical, crown broadening with age, to 20 m or more.

BARK Smooth reddish to grey with horizontal lenticels, sometimes peeling with age.

FLOWERS/FRUIT White flowers 2–3-cm-wide in early spring, before or with emerging leaves; fruit on a single stalk ripens in early summer.

Prunus is a large and varied genus with hundreds of species that include plums, cherries, peaches, apricots and almonds, as well as ornamental flowering cherries and cherry laurels. Some are native to Canada while others such as sweet cherry are introduced, usually for fruit, but some cultivars are ornamentals. Sweet cherries (also called Mazzard cherries) tend to struggle in our climate; they prefer hot summers and cold winters, which explains why the commercial growers are in the Okanagan. Sour or pie cherries make a better bet for a crop here. One of the most popular sweet varieties is Bing, developed in an Oregon orchard and said to be named after Ah Bing, a foreman who worked 35 years for the grower, then returned to China for a visit in 1889 and was not allowed back due to anti-immigration laws.

2535 Venables St W of Kamloops St has a large street tree • several of the Plena cultivar, an ornamental, are in front of the Pacific Railway Station on Main St and Terminal Ave in Thornton Park.

Prunus cerasifera 'Atropurpurea'
purpleleaf plum

LEAVES Elliptical with a pointed tip and toothed, to 10 cm; dark purple to red.

FORM Roundish or vase-shaped, 5–8 m tall.

BARK Dark brown, rougher with age.

FLOWERS/FRUIT White to pink, fragrant and showy, 2–3 cm across, in abundance before leaves emerge; fruit 2–6 cm wide, purple-skinned.

Planted widely as a street tree for the spectacular flower show in spring. Nice as a novelty but a long summer looking at rows of trees with the same dense, dark, light-sucking leaves may not be worth the spring fling. Will produce fruit some summers but they're usually neglected, being the same colour as the leaves and hard to spot, and also because people don't know how tasty they get. Also known as Atropurpurea cherry plum or Pissard plum, named after the French gardener of the Shah who brought it from Persia to Europe in the 1870s.

Street plantings everywhere, just look for the dark purple leaves; across the street from 2358 Cornwall Ave in Kitsilano Park an old gnarled one shows how pretty they can look in isolation.

7 *Prunus domestica*
plum

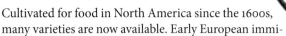

LEAVES Oblong with blunt tip, to 10 cm long.

FORM Small to medium tree with upright branches, to 10 m tall.

BARK Thin, smooth, grey-brown.

FLOWERS/FRUIT Abundant white flowers in early spring.

Cultivated for food in North America since the 1600s, many varieties are now available. Early European immigrants here favoured types such as Italian prune-plums, which may be productive even when conditions are poor. Garden stores now offer more choices of plum trees, along with cherries, on dwarf root stalks, as well as more Asian varieties, so backyard growers may now experiment with a world of fruit.

 NE corner of Killarney St and E 51st Ave.

Prunus emarginata
bitter cherry

7

LEAVES Oval with narrow base, finely toothed; 3–8 cm long.

FORM Slender trunk, ascending branches in narrow crown; to 20 m.

BARK Reddish-brown to grey with rows of horizontal raised slits for gas exchange called lenticels.

FLOWERS/FRUIT Clusters of 5–10 white flowers in spring; fruit less than 1 cm long, red to dark purple, bitter enough to have you remember the name.

Native from BC to Mexico, but not popular in cultivation since the fruit is unlovable and imported flowering cherries show better. Identification in the wild may be complicated by natural hybrids blending this native cherry with introduced sweet cherries. Bitter cherry has medicinal properties with bark extract traditionally used as a blood purifier, laxative, tonic and a treatment for tuberculosis and eczema—but knowing that it contains poisonous cyanide should deter you from casual experimentation. The outer bark, tough and waterproof, can be peeled off in strips to make baskets or decorate tool handles.

Not normally planted as a street tree or in gardens so you'll have to look in forested areas such Stanley Park and Queen Elizabeth Park.

7 *Prunus serotina*
black cherry

LEAVES Oblong to lance shaped with long narrow tip, finely toothed, 5–15 cm long; glossy green, paler beneath.

FORM Tall to 20 m or more with upright branches in the wild; cultivated types vary.

BARK Smooth and grey when young, becoming darker and scaly with age.

FLOWERS/FRUIT Small, white, in terminal spikes 10–12 cm long at the end of new leaf shoots; fruit dark red to black, up to 1 cm in diameter, edible but astringent.

Native to eastern North America where it is common, down to Guatemala; introduced here in various cultivars along streets and in parks. An important timber tree where it grows widely, other names include rum cherry, cabinet cherry and whiskey cherry.

 Five on E 23rd Ave W of Carolina St.

LEAVES Slender with narrow tip, finely toothed (*serulla* is Latin for "little saw"), 5–10 cm long; dark green and staying that colour until they drop in fall.

FORM Smallish tree to 10 m tall with neat, rounded crown and equal spread.

BARK Rich, gleaming, mahogany-red sheets that peel off.

FLOWERS/FRUIT White flowers small and little-noticed among the leaves; fruit small to 1 cm long, red, not deemed edible.

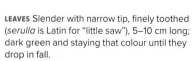

Rare but gaining interest for its bright, glossy, reddish bark which is a standout even in winter when the leaves are bare. The outer layers peel away to reveal a shinier layer beneath. Like many cherries here, though, may not appreciate our wet weather when bacterial canker is rife. Also known as birchbark cherry and red-bark cherry.

 The Sino-Himalayan Garden in the VanDusen Botanical Garden.

LEAVES Oblong to lance-shaped and toothed, often with a long pointed tip; green and sometimes hairy beneath.

FORM Small to medium; varying forms from upright columns to wide-spreading umbrellas.

BARK Usually smooth with horizontal lenticels.

FLOWERS/FRUIT White or pink, from 2–4 cm across, usually borne on a shared stalk and in abundance, often before the leaves emerge so that the crown resembles a white or pink cloud.

Japan sent gifts of cherry trees to Vancouver in the early 1930s. They got planted in Stanley Park. Once they bloomed—whoosh, Vancouver was bound to become a city of *sakura* (Japanese for cherry blossom). The Vancouver Cherry Blossom Festival encourages the cherry madness with guided tours, parties and more. The viewing season may start as early as October with Jugatsu-zakura (*Prunus subhirtella* 'Jugatsu-zakura'). Early spring favourites include Whitcomb (*Prunus subhirtella* 'Whitcomb') and Accolade (*Prunus* 'Accolade'). Yoshino cherries (*Prunus yedoensis*) are famous for *hanami* viewing parties in Japan, with popular

Along streets and in parks all over the city; Oppenheimer Park at Jackson Ave and Powell St has one of the most attractive cultivars, Akebono • check the Vancouver Cherry Blossom Festival website vcbf.ca for blooming dates and event info.

varieties here including Somei-yoshino and Akebono. The village cherry (Sato-zakura) group has Shirotae and the delicately pink-flowered but tough Kanzan which is so widely planted it amounts to more than 12 percent of the city's street trees. With so many types now here you would need a book to keep track—and fortunately there is one, *Ornamental Cherries in Vancouver* by the UBC Botanical Garden's Douglas Justice.

7 *Pyrus calleryana*

callery pear

LEAVES Oval with a narrow tip, to 7.5 cm long; dark green and glossy.

FORM Small to medium tree with upright branching in narrow oval crown to 12 m tall.

BARK Smooth, dark grey when young, maturing to lighter grey with furrows.

FLOWERS/FRUIT Clusters of small white flowers; fruit in small 1-cm-wide balls on long stalks, not tasty like edible pears.

Native to Asia but planted all over North America as a dependable street tree that looks okay and doesn't cause trouble. With many cultivars offering a number of utilitarian benefits to urban designers, some feel it has been overplanted.

 Several in a row of the upright Chanticleer cultivar on the N side of E Broadway from Quebec St to Ontario St, planted in 2007.

Salix x sepulcralis 'Chrysocoma'
golden weeping willow

LEAVES Long, slender, pointed and finely toothed, 4–6 cm long; green above and paler below.

FORM Thick trunk with a broad, weeping crown; to 20 m tall and wide.

BARK Light brown, corky, ridged.

FLOWERS/FRUIT Erect yellow catkins appear with the leaves in spring.

Golden weeping willow is an elegant tree with cascading yellow branches offering grace and colour even in winter. The park-popular beauty is from parents native to western China but carried far beyond their range, beginning with ancient Silk Road traders. It can look great in an open space where the dangling leaves touch the surface of a pond. On the down side, it's fast-growing and short-lived with thick branches of weak wood prone to impressive failure. Still worth planting in a park though.

Parks with ponds all over the city including Trout Lake, Lost Lagoon in Stanley Park and Memorial South Park on 41st Ave at Prince Albert St.

7 *Salix lucida* ssp *lasiandra*
Pacific willow

LEAVES Long, slender, pointed and finely toothed, 5–12 cm long; shiny dark green above, paler below.

FORM Crooked trunk with irregular branches in somewhat rounded crown; large shrub or small tree to 10 m tall.

BARK Dark grey, furrowed.

FLOWERS/FRUIT Catkins appear with leafy shoots, to 12 cm long.

Willow trees can be difficult to sort out with more than 400 species and plenty of hybrids. About 20 (less showy) willows are native to Canada; in our area these include

Hooker's willow (*Salix hookeriana*), Pacific willow (*Salix lucida*), Scouler's willow (*Salix scouleriana*) and Sitka willow (*Salix sitchensis*), all providing indigenous people material for ropes, clothes, medicine and more.

Not typically planted, look for them in natural sites with wet areas such as around Lost Lagoon and Beaver Lake in Stanley Park, or the SW corner of Trimble St and Belmont Ave along with Scouler's willow which has velvety leaves that are widest above the middle area and have rounded tips.

LEAVES Fan-shaped to 1 m in length, divided into linear segments with drooping tips.

FORM Single, short, thick fibrous trunk holding palm-leaf crown; to 12 m tall.

BARK Orange-brown, very rough texture as the fibrous leaf bases remain long after the leaves are gone.

FLOWERS/FRUIT Sprays of small, light, yellow flowers; fruit yellow to blue-black kidney-shaped drupes ripening in mid-autumn.

Not the tree some visitors to Canada expect to see. It could mark another dire climate change warning, but this cold-hardy palm has actually been grown here for decades, especially on warmer sites near the ocean where the coldest winter chills may be softened. Native to China, Burma and north India including the lower Himalayas; also known as Chusan palm or Chinese fan palm.

Beach Ave and Davie St at English Bay where they seem perfectly at home on a warm summer day.

7 *Tilia platyphyllos*
bigleaf linden

LEAVES Heart-shaped with asymmetrical base and long, pointed tip, 12 cm long; green above, paler below.

FORM Medium to large tree 20 m or taller, with upright branching and rounded crown.

BARK Smooth grey when young, becoming darker with furrows.

FLOWERS/FRUIT Small, yellow-green, fragrant flowers in drooping clusters; fruit in small, round nutlets 1 cm or less in diameter.

Linden trees are popular in parks throughout Europe and North America where they make tall and handsome shade trees with fragrant flowers in early summer. They're also popular with aphids, which may make the area beneath the branches a sticky mess. Herbal teas from the blossoms have been used for centuries in Europe to treat ailments from apoplexy to vertigo; the

VanDusen Botanical Garden has a good selection to sort the various species out; Commercial Dr from Napier St to Charles St has opposite rows of *Tilia* 'Euchlora' (Crimean linden), an introduced hybrid planted on many streets here.

dried flowers can also aid sleep when placed in a pillow. There are related species that look similar, some under the common names basswood or lime tree, as well as hybrids between them which may make identification challenging. *Tilia tomentosa* (silver linden) has leaves that are dark green above and white underneath. *Tilia cordata* (littleleaf linden) has smaller leaves that are smoother above and below.

7 *Ulmus americana*
American elm

LEAVES Oval with asymmetrical base and narrow pointed tip, 15–20 veins per side and toothed; dark green above, paler and slightly hairy below.

FORM Vase-shaped with dropping branch tips; thick trunk, often buttressed with age.

BARK Dark brown with shallow intersecting ridges when young, later deeply furrowed and scaly grey.

FLOWERS/FRUIT Small clusters of white flowers on separate stalks; fruit a greenish wafer-like oval samara to 1 cm long.

A majestic shade tree lining streets in countless cities throughout eastern North America, until Dutch elm disease wiped them out. This lesson about the weakness of monocultures is still being learned. Dutch elm disease has not been a problem here, but elms were never heavily planted anyway. It can be difficult to sort out the various species of elms with even experts disagreeing on proper classifications. ID clues to at least put you in the elm camp include rough-textured leaves with an uneven base, irregular teeth, prominent veins and a short stalk. Also look for samaras holding a single seed in a roundish

wafer. The classic shape for an American elm is a tall, graceful vase, making you understand why their loss is still mourned in much of the continent.

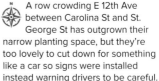

A row crowding E 12th Ave between Carolina St and St. George St has outgrown their narrow planting space, but they're too lovely to cut down for something like a car so signs were installed instead warning drivers to be careful.

8 ▸ *Arbutus menziesii*
arbutus

LEAVES Oblong, thick, leathery, 7–12 cm long; glossy dark green above, paler underneath.

FORM Twisty, leaning trunks and irregular crown, to 30 m tall.

BARK Thin, papery, cinnamon-red, peeling off in curly flakes to reveal smooth yellow-green bark beneath.

FLOWERS/FRUIT Hanging white or pinkish fragrant clusters in spring; fruit in small, orange-red berries.

A gnarled and radiant tree with surreal hues in the shedding bark that would not look out of place on a low budget sci-fi movie set. The only broadleaf evergreen native to Canada; often seen growing out of rocky seaside outcrops where it seems no tree would survive—good drainage is key if you want to try one in your garden. The berries, astringent when raw, may be used to make cider and jellies. The bark and leaves are medicinal, used to treat colds, skin troubles, bladder infections and more. Also called Pacific madrone.

One at the NW corner of Oak St and 37th Ave and a happier one 20 m W of it at the parking lot entrance to VanDusen Botanical Garden.

8 *Asimina triloba*
pawpaw

LEAVES Large, 15–35 cm long, oblong with narrow base and tip, wider above the middle; curious smell when crushed that some liken to cucumbers; green turning to yellow or bronze in fall.

FORM Understory woodland shrub to a smallish tree with an attractive pyramidal shape.

BARK Dark brown, smooth.

FLOWERS/FRUIT Flowers dark purple with three lobes (hence the triloba name), the colour and dank aroma suggest they're meant to attract flies rather than bees; fruit resembles a bent mango, up to 16 cm long, green turning to yellow or brown when ripe; delicious.

The only subtropical fruit native to North America with a range from southern Ontario to Florida and into the midwest. The fruit offers a delectable blend of banana, mango and custard, if you can get one to grow and ripen (cross-pollinating with a different pawpaw cultivar and a hot summer will help).

Strathcona Community Garden has three.

8 *Cydonia oblonga*
quince

LEAVES Oblong to rounded, soft and downy surface; 6–11 cm long, delicate green.

FORM Spreading crown on stout trunk with twisty branches; to 8 m tall.

BARK Thin, smooth, grey, peeling with age.

FLOWERS/FRUIT Big and beautiful whitish-pink blossoms; fruit pear-shaped, ripening to yellow, hard and bitter when raw but cooks into superb jams, jellies, pies.

Native from southwest Asia to Turkey and Iran. This is a charming tree with or without the fragrant fruit, but when it produces it becomes magnificent. You can perfume a room by placing a fresh quince on the table. Plutarch writes of ancient Greek brides eating a quince on their wedding night to sweeten the first kiss. Great choice for a community orchard fruit tree; pilferers get deterred by the astringent flesh that hides its excellent flavour until cooking.

A small one in Grandview Park at Charles St and Commercial Dr, near the Charles St entrance west of the intersection.

8 *Diospyros kaki*
persimmon

LEAVES Oval with pointed tip, to 20 cm; glossy yellowish-green turning to yellow in fall.

FORM Single or multi-stemmed, small to medium sized tree with attractive, dense, spreading branches.

BARK Grey and scaly.

FLOWERS/FRUIT Small yellow flowers in spring; fruit varies with the many cultivars (said to be more than 2,000 in Asia), but the kaki variety grown here is orange, thick skinned and, if you're lucky enough to get one to ripen, arrestingly sweet.

North America has a native persimmon (*Diospyros virginiana*) but the Asian varieties are gaining popularity here because the fruit is better. Vancouver summers (to date) are usually not long and hot enough for a reliable crop, but the tree is attractive and may merit planting even if the yield is only sporadically successful.

Several in the Strathcona Community Gardens at Prior St and Hawks Ave.

Magnolia grandiflora 8
evergreen magnolia

LEAVES Broadly oblong with narrower base and tip, thick and leathery, to 15 cm long; glossy dark green.

FORM Pyramidal to rounded crown.

BARK Smooth grey, some scales with age.

FLOWERS/FRUIT Creamy white flowers 20 cm wide with a fine lemony fragrance when they bloom, which isn't every year in our climate; fruit in 7- to 10-cm-long reddish-brown fuzzy aggregates of seeds.

Native of the Deep South in the US, yet able to survive and stay green through our mostly clement winters. Also known as southern magnolia, they offer an oddly tropical look to gardens here with their lush growth in winter and exotic flowers. Both are a plus, but the downside for some is the dense shade they cast.

A grove of 7 or 8 at the entrance to the outdoor pool at Kitsilano • Dr. Sun Yat-Sen Classical Chinese Garden at 578 Carrall St.

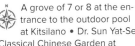

8 *Magnolia x soulangiana*
saucer magnolia

LEAVES Oblong, stiff, arched with wavy edge, to 12 cm long; dark green above, paler beneath.

FORM Short-trunked, often multi-stemmed spreading tree to 10 m tall.

BARK Smooth, pale grey.

FLOWERS/FRUIT Large, pinkish-white, tulip-like blooms appearing on bare branches in spring.

Magnolia trees come in hundreds of varieties, some of which are blends, making precise identification difficult. You can usually recognize the genus by big, showy flowers and unusual cob-like seed-cone fruit. Saucer magnolia, also called soulange magnolia, is a hybrid cross between two Chinese magnolias by a French gardener in 1820. It was admired for its stunning floral display in early spring before the leaves emerged, and thus planted widely everywhere including Vancouver. Kobus magnolia, a Japanese introduction, has gained popularity here as a street tree, while gardeners' choices for smaller spaces include the shrub or small tree-like Stellata cultivar grown for its brilliant white star flowers.

Two saucer magnolias in front of the planetarium in Vanier Park were originally planted in 1912 along Georgia St; due to be cut down for construction in 1968, they were saved and re-planted here • Culloden St west of Knight St between 41st Ave and 43rd Ave has a dozen kobus magnolias of various sizes.

LEAVES Pointed oval on a short stalk, 10 cm long; glossy dark green above and paler beneath, turning to radiant yellows and reds in fall.

FORM Strong pyramidal form when young, maturing to rounded or irregular shape, 15–20 m tall.

BARK Grey with rectangular ridges, deeply furrowed with age.

FLOWERS/FRUIT Blue-black berries 1–2 cm long, hanging on long stalks; may be eaten but they're sour.

Native to eastern North America in open forests and the edges of swamps. Tried as a street tree here fairly recently with glowing success so they're becoming more popular. The fall colours are spectacular enough to make anyone want one. Also known as black gum or sour gum.

In 1997 four were planted in a row on E 15th Ave W of Woodland Dr across from Clark Park.

8 *Oxydendron arboreum*
sourwood

LEAVES Narrow oblong tapering to a pointed tip, 12–20 cm long; green above, paler beneath, lovely fall scarlets and oranges.

FORM Pyramid shape with ascending branches; to 12 m or taller.

BARK Grey, tinged with red, ragged furrows that may reveal orange beneath.

FLOWERS/FRUIT Long clusters of nodding, creamy white, bell-shaped flowers; fruit ripens to small, upright, five-angled capsules along spikes which stay on the tree into winter.

Native to southeastern US but planted elsewhere for attractive flowers and gorgeous fall colours. A mature specimen makes a fantastic tree even without the autumn display. Also known as sorrel tree or lily-of-the-valley tree.

Hudson St at the Crescent (Shaughnessy Park) just W of the sugar maple near the Osler St entrance.

Stewartia pseudocamellia
Japanese stewartia

LEAVES Oblong and pointed, finely toothed; dull green above, paler beneath.

FORM Smallish tree with single or multiple stems, to 12 m.

BARK As a bonus for an already attractive tree, the bark exfoliates in interesting strips of orange, grey and brown.

FLOWERS/FRUIT Large white flowers look like camellias, thus the *pseudo* (false) and *camellia* in the name, unless you think they look more like eggs sunny side up; brown pointed seed pods remain on branches in winter.

Native to China and Japan, becoming more popular here as a well-shaped tree with pretty flowers and interesting bark that can fit into most urban spaces. Stewartia, like camellia, are members of the tea family or Theaceae.

Eight small trees were planted on School Ave W of Kerr St in 2013.

8 *Styrax japonicus*
Japanese snowbell

LEAVES Oblong with a pointed tip, to 7 cm long; glossy green.

FORM Small pyramidal tree with knobby trunk and crown that spreads with age.

BARK Smooth dark grey when young, turning rougher with age.

FLOWERS/FRUIT Small, white, fragrant, bell-shaped flowers hang down from the branches in late spring; fruit hangs in small, grey-green balls.

A fine patio tree, especially if you sit underneath the graceful branches and look up into the dangling floral display. With the trend towards smaller, manageable, pretty trees for increasingly compact urban spaces, Styrax have become popular, and deservedly so. Less seen but worth checking out is its cousin fragrant snowbell (*Styrax obassia*) which has bigger, bolder leaves and light brown, inedible fruit balls.

Six were planted in 2002 on Nootka St between E 5th Ave and E 6th Ave • a row of fragrant snowbell can be seen in the Yaletown Park planters at Mainland St and Nelson St.

Ailanthus altissima—tree of heaven

Ailanthus altissima
tree of heaven

LEAVES Large with 11–41 leaflets, each narrow, sharply pointed, often toothed on one side near the base unlike the similar compound-leafed walnut; up to 75 cm long; unpleasant smell when crushed.

FORM Straight or crooked trunk, often forked; thick and upright branches in an open, broad crown to 25 m or taller.

BARK Smooth, greenish-grey with prominent branch scars, becoming rougher with pale, vertical lines when older.

FLOWERS/FRUIT Small, yellowish-green in clusters at shoot tips; seed cases in dense clusters of yellowish, hanging, twisted wings.

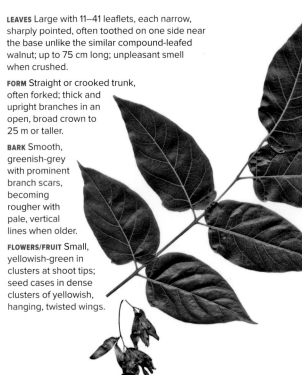

Tree of heaven may have been an ambitious name, given its reputation in much of the continent. Introduced from China to North American gardens in the 18th century, it escaped cultivation to become an invasive species in eastern areas where its indifference to pollution, poor soils and lack of irrigation helped it spread. It is less aggressive here, where a lone tree may add visual appeal to a landscape. Also known as Chinese sumac or ghetto palm.

The odd triangular park where Kingsway, Fraser St and 15th Ave come together • also in Thornton Park at Main St and Terminal Ave.

9 *Albizia julibrissin*
mimosa

LEAVES Large and lacy, made of 8–16 leaflets, each with tiny sub-leaflets; to 20 cm long.

FORM Single or multi-trunked, low-branching, spreading, flat-topped; to 10 m tall.

BARK Smooth, light brown to grey.

FLOWERS/FRUIT Powder-puff clouds of pink in mid to late summer; fruit in flat, green to brown, cross-banded, 15-cm-long pods that hang on the tree through winter.

A tropical looking tree with delicate leaves and light pink flowers all seeming too dainty for Canada. Native to a wide region from Iran to Japan, it's considered invasive in eastern North America but hasn't tried to take over territory here. Reported to be short lived at just 10–20 years, but that may be wishful thinking from those where the tree is considered more of a problem, or where a widespread vascular wilt disease has reduced their numbers. Has a curious habit of failing to leaf out some

years, yet surviving to do fine the following year. Also known as silk tree.

 VanDusen Botanical Garden has four, including the cultivars E.H. Wilson, Red Silk and Summer Chocolate.

Gleditsia triancanthos
honey locust

LEAVES Delicate with 14–30 leaflets on a stalk 15–20 cm long; leaflets with rounded tips.

FORM Short trunk leading to multi-branches in a broad, open, flat-topped crown, to 30 m tall.

BARK Smooth, light brown when young, maturing to grey with deep furrows and scales.

FLOWERS/FRUIT Small greenish-white flowers; seed pods brown, flat, twisted, 30–40 cm long.

Native to eastern and central North America but widely planted elsewhere, including here, as an attractive but tough tree that casts a delicate light shade. *Triancanthos* is from "three" and "thorns" to describe the very thorny standard, but the planted trees here are typically one of several thornless cultivars.

The N side of W 37th Ave from Oak St to Willow St has a row of 13 Skyline honey locust planted in 1996.

9 *Juglans cinerea*
butternut

LEAVES Large compound leaves with 11–17 leaflets, 30–60 cm long; terminal leaflet usually same size as nearby leaflets while lateral leaflets diminish in size towards the leaf base; yellow-green above, pale and hairy underneath.

FORM Short, forked trunk leading to open, irregular crown to 25 m tall.

BARK Smooth and grey when young, maturing with dark, intersecting fissures.

FLOWERS/FRUIT Pollen flowers drooping in yellow-green catkins 6–14 cm long; fruit in elongated, pointed 5–8-cm-long green husks with a sticky, hairy surface.

Native to eastern North America from New Brunswick to Arkansas. More valued for nuts than the wood, the butternut name refers to the buttery taste. Faster growing and shorter lived than black or English walnut. Like others in the Juglandaceae family, it emits a chemical called juglone that turns the soil beneath it toxic to competing plants. Butternut come in several cultivars

In front of 2196 Yew St at 37th Ave with several more along Yew St up to 40th Ave • VanDusen Botanical Garden has an attractive one next to a black walnut, both NW of the visitors' centre.

which may make identification confusing; it may also hybridize with English walnut to produce a tree that makes classification even more challenging. Recall that English walnut leaves have fewer (5–9) leaflets and the familiar nuts have thin shells easy to crack. Also known as white walnut.

9 *Juglans nigra*
black walnut

LEAVES 15–23 leaflets on a central stalk, 20–60 cm long; middle leaflets larger than either end, terminal leaflet smaller than the rest or missing; yellow-green above, slightly hairy below.

FORM Straight trunk, open rounded crown of thick branches, to 30 m tall.

BARK Light brown, scaly, maturing with dark, intersecting ridges.

FLOWERS/FRUIT Pollen flowers in catkins 5–10 cm long; seed flowers in erect clusters of 1–4 flowers; fruit in 4–6-cm-diameter globes with deep grooves, may be in clusters of 2–3 nuts.

Native to eastern North America where it was almost logged out for its fine, durable wood. Nuts produce a bothersome stain and are tough to crack, but delicious if you do. Black walnuts have a long history for nobility, and to see a huge and graceful one is to understand why.

 In front of 3544 E 28th Ave, E of Skeena St, a 30 m tall stupendous black walnut rules over the street.

9 *Juglans regia*
English walnut

LEAVES 5–9 leaflets with terminal leaflets largest; blunt and wide with smooth edges.

FORM Large tree to 40 m tall with long spreading branches on a thick trunk.

BARK Grey and smooth when young, rougher later.

FLOWERS/FRUIT Hanging green catkins to 12 cm long; familiar nuts in green, smooth husks that partially split open when ripe.

Walnuts are a more complicated genus than just this familiar type we buy in the stores. There are Japanese walnuts (*Juglans ailantifolia*), which have very large leaves of 13–17 bright green leaflets, Manchurian walnuts, northern California black walnuts, and hybrids to boot. *Juglans regia*, also known as Persian walnut, common walnut and royal walnut, are a global favourite once rightly deemed fit for a king.

 In front of 3642 Pt. Grey Rd a 30 m tall English walnut demonstrates their height and spread given enough time • also behind 2099 Beach Ave.

Laburnum x watereri

golden-chain tree

9

LEAVES Three pointed ovals each about 7 cm long that fold up at night.

FORM Large shrub or small tree, often grafted onto common laburnum rootstock and pruned into a weeping tree.

BARK Greenish-brown.

FLOWERS/FRUIT Chains of golden pea-shaped flowers 20–30 cm long; seeds in flat, brown pea-pod cases remaining on the tree into winter.

Planted chiefly for its dazzling flower display, it has earned its reputation for splendour. Less mentioned is the fact all parts of the tree are poisonous, but since no one seems tempted to eat them it hasn't been a problem here. The Vossi cultivar planted at the Van-Dusen Botanical Garden has longer flower chains than the standard, and since that's pretty much the point of this tree it is justifiably popular when blooming.

 Best place to see it is the Laburnum Walk in VanDusen Botanical Garden when it draws crowds with a spectacular flower show in spring.

9 *Pterocarya fraxinifolia*

Caucasian wingnut

LEAVES 11–27 leaflets on a single stalk, broad and floppy, to 60 cm long.

FORM Short, thick trunk leading to a spreading crown, to 30 m tall.

BARK Dull grey with broad, vertical ridges.

FLOWERS/FRUIT Female catkins in chains 25–50 cm long; seeds surrounded by yellowish-green papery wings.

Not commonly planted as a street tree due to shallow roots and a wide-spreading crown, but the colossus in the West End mentioned below is a neighbourhood feature. Worth visiting just to see a tree that celebrates its own magnificence with flowery tassels. Native to the Caucasus region of eastern Turkey and northern Iran, if it wasn't so imposing it would surely get planted in more yards and gardens. The related species sometimes seen here is the similar Chinese wingnut (*Pterocarya stenoptera*), differing in having leaves with a central stalk that's winged and with nutlets longer and narrower than those of Caucasian wingnut.

 The SW corner of Chilco St and Comox St.

9 ▶ *Robinia pseudoacacia*
black locust

LEAVES 7–19 oval leaflets including a terminal leaflet on a central stalk 20–30 cm long; two spines at the base of each leaf; leaflets oval with bristling tips, 3–5 cm long; dull green.

FORM Medium-sized tree with irregular trunk and short gnarled branches, the form makes it a visual treat even in winter.

BARK Smooth and brown-grey when young, maturing to deeply furrowed dark brown.

FLOWERS/FRUIT Fragrant white pea-like flowers in long 10–14 cm clusters in early summer; seed pods flat, brown, 7–10 cm long and lasting through winter.

With seeds widely spread by birds and animals, and a capacity to thrive even in poor soils, black locusts are considered weed trees in much of North America. They can be fast-spreading here as well, but may also improve an otherwise dull landscape. They look like they would be at home on the African savannah. As a member of the bean family (Fabaceae) they have nitrogen-fixing bacteria on their roots, making them early adopters improving poor soil. The wood is very strong—early settlers used it to make nails. Frisia is a popular

cultivar with yellow leaves that seem to glow, sometimes offering the only visual joy to a grey day. The purple-flowered *Robinia* 'Purple Robe', which some experts now say shouldn't be considered a *psuedo-acacia* species, is getting more popular as a durable adornment.

Kitsilano Beach Park where Creelman Ave ends at Arbutus Blvd • you can see Frisia on 8th Ave from Scotia St to Prince Edward St • Quebec St between E 55th Ave and E 56th Ave has several Purple Robe black locust mixed with some Halka honey locust (*Gleditsia triancanthos* 'Halka').

9 *Rhus typhina*
staghorn sumac

LEAVES 11–31 leaflets on a central stalk 30–50 cm long; leaflets narrow, sharply pointed, 5–12 cm long and sharply toothed; green turning to bright red or orange in fall.

FORM Multi-trunked shrubs or small trees to 6 m tall.

BARK Thin, grey-brown when young with prominent lenticels, growing scaly with maturity.

FLOWERS/FRUIT Small, greenish-yellow in large clusters at shoot tips; fruit small, red, in cone-shaped upright clusters, may be crushed into a refreshing drink.

The common name is derived from the twiggy branches covered in soft hair, somewhat like a stag's antlers. By contrast, smooth sumac (*Rhus glabra*) have smooth and hairless twigs. Fruit on female trees ripens into striking red clusters that last into winter when birds may appreciate a meal. Sumac seem to find their own way into (ecologically) rough neighbourhoods and back alleys, and thus get ignored, but the vivid fall colours make them impossible to miss.

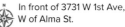
In front of 3731 W 1st Ave, W of Alma St.

LEAVES 9–17 leaflets, each 3–5 cm long, blunt-tipped and toothed.

FORM Smallish trees with light, open crown, to 15 m tall.

BARK Smooth, may be rougher and peeling with age.

FLOWERS/FRUIT Flowers on hairy stalks; small acidic berries eaten by birds and animals including some humans who process them into jam or jelly.

A member of the rose family (Roseacea), so despite the name they have nothing to do with ash trees, although the compound leaves are reminiscent. The fruit is said to be an important avian food source. Then they seem to remain on the tree so long into winter you have to wonder if birds got the memo. Also known as rowan, and an important tree in mythology with deep meaning from the medicinal to the spiritual for cultures in many lands. The centre of the Traditional Medicine Wheel Healing Garden in Oppenheimer Park is an oakleaf mountain ash (*Sorbus x thuringiaca* 'Fastigiata').

 Waterloo St between W 13th Ave and W 15th Ave has nine mountain ash trees • in front of 818 E Pender St E of Hawks Ave is the Cardinal Royal cultivar which puts out more berries than the standard.

With trees from everywhere and Stanley Park defining downtown, Vancouver is the tree geek capital of the world.

Bibliography

Ancient Trees. Edward Parker & Anna Lewington. Batsford. 2013.

Aristocrats of the Trees. Ernest H. Wilson. The Stratford Company. 1930.

Cedar. Hillary Stewart. Douglas & McIntyre. 1984.

Identification Guide to the Trees of Canada. Jean Lauriault. National Museum of Natural Sciences. 1989.

Hiking Guide to the Big Trees of Southwestern British Columbia. Randy Stoltmann. Western Canada Wilderness Committee. 1987.

Ornamental Cherries in Vancouver. Douglas Justice. UBC Botanical Garden and the Vancouver Cherry Blossom Festival. 2014.

Manual of Woody Landscape Plants. Michael Dirr. Stipes Publishing Co. 1990.

The Meaning of Trees. Fred Hageneder. Chronicle Books. 2005.

North American Landscape Trees. Arthur Lee Jacobson. Ten Speed Press. 1996.

New York City Trees. Edward Sibley Barnard. Columbia University Press. 2002.

Plants of the Northwest Pacific Coast. Jim Pojar and Andy MacKinnon. Lone Pine. 1994.

Sibley Guide to Trees. David Allen Sibley. Alfred A. Knopf. 2009.

Trees of Greater Victoria: A Heritage. G.H. Chaster and others. Heritage Tree Book Society. 1988.

Trees in Canada. John Laird Farrar. Fitzhenry & Whiteside. 1995.

Trees of North America. C. Frank Brockman. Golden Books. 1986.

Trees of the Northwest. J. Duane Sept. Calypso Publishing. 2011.

Trees of Vancouver. Gerald B. Straley. UBC Press. 1992.
Vancouver's Heritage Tree Inventory. Elisabeth Whitelaw
 and Clarence Sihoe. Privately published through the
 BC Heritage Trust. 1983.

*Acer macrophyllum—*bigleaf maple

Glossary

allee: a walkway lined on either side by trees.

angiosperm: flowering plant; the deciduous trees in this guide are angiosperms.

bract: modified leaf at the base of a flower or cone.

cambium: thin layer of living cells that develop outward into phloem (bark) or inward into xylem (wood).

conifer: cone-bearing tree.

cultivar: the culitvated variety of a species, typically grown for some unique feature(s) that make propagating worth it.

deciduous: shedding or falling off at the end of a season, such as trees which drop their leaves in fall.

genus: a group of related species of organisms.

glauca: blueish grey.

glabrous: smooth, as in a leaf or stem without hair or down.

grove: a small group of trees, usually without understory vegetation.

gymnosperm: non-flowering seed plant; the conifers in this guide are gymnosperms.

lenticel: a pore, often horizontal in shape, in the bark of a tree allowing gas exchanges.

lobe: leaf or petal segment, divided from other segments by the sinus.

native plant: occurring naturally in an area without having been imported.

naturalized: a non-native which has "escaped" cultivation and can now grow and reproduce itself in the wild.

peduncle: stalk of a leaf.

scientific name: a botanical classification system started by Linnaeus in 1753 in his book *Species Planatarum* which grouped plants by assigning each to a genus (such as *Thuja*) and a more specific species (such as *plicata*); sometimes called the "Latin name" although incorrectly since Greek and other words may be used instead.

species: a group of related organisms which can inter-breed.

sinus: the space between lobes in a leaf.

stomata: a pore in a leaf or stem allowing gas exchanges.

syconium: a fleshy, hollow receptacle that contains numerous fruitlets such as a fig.

toothed: the edges of the leaves are not smooth but have saw-like points; double-toothed means the teeth themselves also have teeth.

Magnolia dawsoniana 'Chyverton Red'—Chyverton Red magnolia

Acknowledgements

Vancouver Tree Book is in your hands thanks to the diligence and kindness of many, including some I have never met.

Plant researcher and UBC Botanical Garden curator of collections Gerald Straley wrote *Trees of Vancouver* in 1992. I found it an invaluable resource for identifying trees in the city. Sadly, Straley died just five years later and the book is out of print. This work is an attempt to stand on his shoulders.

On a reporting assignment to New York City I met a designer who had drawn a map of all the trees in Central Park. When I asked where he got the wacky idea, he mentioned Edward Barnard, author of *New York City Trees*. I found the book perfectly sized for urban tree explorers, and contacted Barnard to ask how he did it. True to some unspoken code among tree people to help anyone follow the passion, he was generous in sharing details of the publishing process.

In starting to research this book, I had to face the fact I'm no marvel when it comes to tree identification. I have an arborist background but my work never required knowing *all* the trees in the urban forest. And there's so many of them, right? Fortunately I knew some experts whose vast knowledge puts them at a level so lofty I can only gaze up in wonder at how they got there.

Douglas Justice, assistant director and curator of collections at UBC Botanical Garden, has an encyclopedic knowledge of trees matched by a gift for sharing it. He taught the tree ID course while I was learning landscape architecture at UBC, and it helped shape my views on which trees work best in our bioregion. His *Vancouver Trees* app has details on more trees that you probably knew existed.

Bill Stephen, arborist for the City of Vancouver, is one reason the urban forest is doing so well. He was generous with information on city strategies for street and park trees, the locations of certain trees mapped in this book, and background on the history of how we got the forest we have. He's a true friend of the trees.

Alex Downie, also from the city's Park Board that manages trees, helped locate trees. Marina Princz, librarian at the VanDusen Botanical Garden, aided with research at her site. She also marshalled others from the garden, including Samanta Sivertz, who writes the garden's lively tree profile brochures and helped with the map, and volunteers Erica Dunn and Pam Frost who offered advice on which trees to include in it.

Wendy Cutler is a Vancouver volunteer-extraordinaire for worthy tree projects. Her years of tracking down and photographing interesting trees were put to good use. Others helped in countless ways, from field work to publishing advice to expert interviews to sharing enthusiasm for trees and a book that would help others appreciate them too. These included Hartley Rosen, Yolanda Bienz, Ira Sutherland, John Worrell, Celia Brauer, Renee Sarojini, Tim Louis, Richard Mackie, Susan Safyan, the staff and volunteers from Tree City and the Environmental Youth Alliance and everyone else who helped make the TreeKeepers project great.

Fraxinus pennsylvanica—green ash

Index

Bold numbers are the main profile page(s) for that tree.

Allee of *Acer saccharinum*—silver maple